The Student Supercomputer Challenge Guide

ASC Community

The Student Supercomputer Challenge Guide

From Supercomputing Competition to the Next HPC Generation

ASC Community
Inspur (Beijing) Electronic Information
Industry Co., Ltd
Beijing
China

ISBN 978-981-13-3831-1 ISBN 978-981-10-3731-3 (eBook)
https://doi.org/10.1007/978-981-10-3731-3

Jointly published with Science Press, Beijing, China

The print edition is not for sale in China Mainland. Customers from China Mainland please order the print book from: Science Press, Beijing, China.
ISBN of the China Mainland edition: 978-7-03-047163-5

Printed on acid-free paper

This Springer imprint is published by the registered company Springer Nature Singapore Pte Ltd.
part of Springer Nature
The registered company address is: 152 Beach Road, #21-01/04 Gateway East, Singapore 189721, Singapore

Foreword

Nowadays, supercomputers are widely used in areas such as weather prediction, geophysics exploration, earth modeling, developments of new material and drug, as well as automobile and aircraft designs. In the meantime, supercomputing is also an important method for dealing with essential social and economic issues. For example, decisions can be made faster and more effective in environmental protection, gene sequencing, natural disaster prediction and Internet deep learning with the help of supercomputing.

Recent years have witnessed the fast development in building more powerful supercomputers. The computing performance of top supercomputers has reached to over 50 Petaflops (PFLOPS), and will be soon approaching into Exaflops-scale era in future years. Despite of the fast development of supercomputing hardware, and the promising trend of the supercomputing industry, we see a gap between the system and application, as well as the lack of technical talents.

In 2012, the first Asia Supercomputing (ASC) Competition was organized by Inspur and the Asia Supercomputer Community to provide a wider platform for gathering and fostering young students around the world. It was a good opportunity for more young students to understand the latest supercomputing techniques and applications, and to have systematical lessons and practical operations. In addition, such competition can promote the passion for young students in supercomputing innovations. Based on ASC, Inspur hopes to make some contributions to the development of supercomputing, which is also a responsibility for commercial companies like Inspur.

We are delighted to see that ASC is gaining a positive response and support from more and more universities, research institutes, and supercomputing experts. Along with the Supercomputing Conference (SC) in the USA and the International Supercomputing Conference (ISC) in Germany, the ASC has become one of the world's top three supercomputing competitions. The advanced technology, high-end system, real-world application, and the practical operations, have encouraged more and more young students to express their passion, imagination, and innovation. Those students have achieved inspiring results in the competition, which in return promotes the development of real application. Furthermore, with more and more students being interested in ASC and choosing supercomputing as their future career, it is an important step to help the future development of supercomputing.

As the first book that systematically introduces supercomputing competitions, the completion of this book contains efforts from many student teams and supercomputing experts. It contains essential knowledge and provides an easy start. Making supercomputing more general and popular is one of the goals of ASC, so that we hope this book publication will be of help to young researchers in supercomputing.

December 2015

Endong Wang
Academicians of Chinese Academy of Engineering
Director of State Key Laboratory of High-end
Server & Storage Technology
Chief Scientist of Inspur

Together with theory and experiment, computation has become an important solution for scientific research and engineering design. Numerical simulation based on supercomputing has been widely used in essential areas relating to national economy and life science, such as fission and fusion energy, development of new energy, design of new material, weather prediction, electromagnetism information safety, geophysics exploration, structure design of aircraft and high-speed train, and simulation of large-scale engineering. As a result, both performance and accuracy are improved, while the cost is decreased. Nowadays, supercomputing techniques are much related to human life, and can be found everywhere, such as in accurate weather prediction, simulation of new drugs, disaster precisions, environmental protections, large-scale information services, data mining, and digital media and cultures.

The development of supercomputing in China is still slower than in some leading countries, so that more talented young students and researchers are in great demand in order to mine the supercomputing potential in solving real problems. Therefore, ASC was born under the organization of the Asia Supercomputer Community and Inspur, and has achieved positive response and huge influences. This book is written by related researchers and competition teams that have been involved in previous ASCs, and will help attract more participates. Through introducing the knowledge and theory corresponding to supercomputing, this book will push forward the development the supercomputing development of our country, absorb and foster more young researchers.

This book will help readers understand supercomputing and corresponding system architecture and evaluating methods, know about the ASC competition and its rules, and see the ASC contents, analysis, and optimizing techniques. This book could not have been completed without the contributions of the Asia Supercomputer Community, and Inspur, which is a leading technical company that mainly focuses on important areas such as High-performance Computing (HPC), fault-tolerant computer, big data storage, cloud computing and large-scale application software.

It was a great honor for me to serve as chair of the review committee, and as one member of the supercomputing community. I hope the ASC continues to develop, all the teams achieve further inspiring results, and the publication of this book is rewarded.

March 2015

Zeyao Mo
Deputy director of Institute of Applied Physics
and Computational Mathematics
Chair of the ASC Review Committee

Together with experimental observation and theoretical analysis, computer science and engineering has become an important research method to push forward innovation in engineering and technology. Computer science and engineering allows quantified and accurate descriptions, and is essentially important to improve the technology level. In the twenty-first century, interdiscipline is the most common mode for computer science and engineering, and is helping generate the greatest scientific breakthroughs. Supercomputing is one of the indispensable tools in computer science and engineering.

Recent decades have seen the fast development of supercomputing. Developed countries such as the USA, Japan, and EU countries have invested heavily in the development of supercomputers and implement-related computing techniques and application studies. In China, supercomputing has been promoted to a strategic technique and has been written into the mid-term and long-term development program. Supercomputing has long received support from the 863 program and a lot of achievements have been made. Sponsored by the 863 program, the Tianhe-2 supercomputer has been the most powerful supercomputer in TOP 500, since its emergence in June 2013. Compared with the USA that has the leading supercomputing techniques, there is still a long way to go for China in terms of both the research and applications. Therefore, it has become an urgent demand for China to improve the supercomputing techniques.

The ASC, organized by Asia Supercomputer Community and Inspur, is especially meaningful in order to popularize the supercomputing technology. ASC has gained wide support from a lot of universities all over the world. More and more Chinese universities start to participate in the ASC so that their students can better understand supercomputing, become interested into it, and even devote themselves to its development. Therefore, a book that introduces the basic methods, basic technology and basic rules of supercomputing, as well as explains some information about the competition, is in great demand.

To meet the above demands in supercomputing competitions, this book was published, with efforts from many supercomputing reviewers, advisers, and distinguished scholars. It contains three parts. The Study part is the basic part that aims to help readers to understand the basic conceptions and methods of supercomputers, including the system architecture, network communication, development environment, application environment, and evaluation techniques. The Competition part introduces and analyzes the rules of ASC to make it more transparent. The Advances part analyzes the questions of ASC finals, and provides corresponding suggestions, especially for some special parallel and optimizing methods that target specific applications. This book is aimed to be helpful to students to grasp necessary parallel methods and optimizing strategies in a short time.

The publication of this book will push forward the popularity of supercomputing, and foster more talented researchers. My best wishes that this book can be meaningful for the teams to optimize their applications, and that the future ASCs will be increasingly successful.

July 2015

Guangwen Yang
Director of the Institute of High Performance
Computing at Tsinghua University
Chair of the ASC 2012 Organization Committee

The development of China now faces a series of big challenges, such as the extreme weather disasters, environmental pollution, infectious diseases, a lack of resources, and an extensive inefficient development mode and economy. Therefore, an environment-friendly, highly efficient, healthy and sustainable way of development is in great demand. HPC ought to play an indispensable role during the development.

Under the support of 863 projects, HPC development in China has been greatly promoted in the past 20 years. Supercomputers made in China have achieved the top computing performance in TOP 500 on many occasions. However, we have to admit the large gap in both the breadth and depth, compared with the developed countries. Among many factors that restrict the development of HPC, the shortage of talented people that are skilled at using computing tools to solve real problems, is one of the most important.

ASC is a good trial to foster talented young people in HPC. The competition can improve the sense of competition of the students, foster their abilities using computers to solve real problems, and train them in the skill of optimizing applications with certain limitations. Such experiences will have long-term influences on the careers of those students. Therefore, ASC has gained wide acceptance and favors from those students. The improving number of participants not only contains teams from China, but also contains teams from other countries. ASC has become an influential international activity in HPC.

As the guidance of ASC, this book explicitly introduces the concepts of supercomputing, the competition rules and platforms, and some typical cases and optimizing methods. This is a textbook that can help college students and team advisors to understand and grasp ASC. The publication of such a book will further improve the influence of ASC, encourage more students to participate in the competition, and push forward the advances of HPC techniques and applications.

June 2015

Depei Qian
Chief Scientist for the 863 Key Project on High Performance
Computer and Core Software
Professor of Beihang University
Member of the ASC Advisory Committee

With the fast development of information technology, HPC has become a frontier technique that is widely used in scientific study, geophysics exploration, climate prediction, astronomy and national defense, biology and gene, and new applications in economy, government, industry, and the Internet. However, even though China has achieved world-leading achievements on studying the systematic architectures in supercomputers it is still not qualified as a strong country in HPC due to the lack of people with an interdisciplinary background, and software that is developed independently.

Talent training is one of the important factors to help the development of HPC. The ASC, co-organized by Asia Supercomputer Community and Inspur, is a good platform to train and foster young talents in HPC. Such a platform has covered areas from all over the world and effectively cultivated the interests and passions of students in HPC. It is of great importance to help foster and find the talented people for HPC.

The publication of this book can help more university students to understand HPC and the ASC, and improve their ability and skills. It can push forward the development of the HPC industry in China, and provide future forces for HPC.

Weimin Zheng
Professor of Tsinghua University
Executive Chair of the China Computer Federation
Member of the ASC Advisory Committee

Computing is an important method to help human explore the world's unknown things. As the research targets become more complex, and the continuous pursuit for higher accuracy to understand the world, computing power will be in large demand. Nowadays, the standard of HPC can represent the technology level of a country. History has proved the indispensable value of HPC in scientific study, national economy and social development. As one of the major consumers of HPC, geophysics exploration has a long-established tight relationship with HPC, and so will it be in the future.

As a national enterprise, Inspur is committed to focusing on application innovation and talent cultivation, and puts up a great investment to host ASC. It not only advertises the brand, promotes the popularity of supercomputing and the communication between different countries, but also provides an outstanding platform for enterprises to present HPC problems, and to discover talented young people in HPC.

My sincere wishes to Inspur and I hope it can make more contributions to push forward the development of science and society; and to ASC, so that it can further promote the cultivation of HPC talent.

June 2015

Guoan Luo
Secretary of R&D center of BGP
Member of ASC Evaluation Committee

To deal with the challenging problems during the progress of human civilization, supercomputing has drawn more attention to and played am indispensable role in combating challenges and realizing innovations. Another focus is on the cultivation of talents in supercomputing development and application. The foundation and development of ASC manages to provide an important platform to promote the technical communication of supercomputing technique, to improve the standard of supercomputing applications, and to facilitate the cultivation of supercomputing young talent. In only four years, ASC has helped improve the interests, vision, and the ability of Chinese students in supercomputing.

The publication of this book will help further popularize the basic information of supercomputing, train the competition technique in supercomputing, and improve the level of supercomputing application. In addition, it will also help absorb more college students into learning and participating in the competition, and guide them on the way to become outstanding supercomputing experts. It is certain that through continuous struggle we are able to not only develop the world's top supercomputers but also lead the development of supercomputing technology.

June 2015

Yizhong Gu
Director of the Network Information Center
Shanghai Jiao Tong University
Chair of the ASC13 General Organization Committee

Solving the essential problems in scientific engineering through supercomputers requires not only the HPC hardware system, but also corresponding software to fully explore the performance. The key issue falls into the talented individuals with various supercomputing skills. As ASC encourages the innovation of supercomputing, it will promote the progress of educating more talented people in supercomputing.

ASC focuses on the development of supercomputing and the cultivation of young talent. It encourages students to finish the system ensemble, architecture transferring, parallel computing and performance optimization by themselves, and makes the competition become a field battle for educating supercomputing skills. Therefore, it effectively eases the unbalanced situation between the developments of software and hardware, and serves as the infrastructure for technical innovation in supercomputing.

It is no doubt that a better future of supercomputing for China will rely on the younger generations, as the proverb says 'Better younger generation makes a better nation'.

June 2015

Yuesheng Xu
National Distinguished Scholar of China (Qian Ren Ji Hua)
Guohua Chair Professor of Sun Yat-sen University
Chair of the ASC13 General Organization Committee

In the past four ASC competitions, I was very honored to work in ASC 13 and ASC 14, and serve as the reviewer in the presentation committee in the ASC finals. ASC is an excellent competition that presents us with individual wisdom and teamwork. I was very impressed by the passion, the teamwork, and the spirit of innovation. ASC can help promote the cultivation of more talent for our country, and the innovation in supercomputing applications. Nowadays, ASC has become an important platform to generate more young leaders in supercomputing.

One of the test cases for ASC14 final was the climate ocean model, LICOM. LICOM is developed by our lab, the IAP/LASG, and will be the optimizing targets for the competitors on Tianh-2, the world's top supercomputer. I was very delighted and inspired by the variety of optimizing proposals. LICOM is the key component of the climate model developed by IAP/LASG. While the competition for developing better climate models becomes more and more serious, ASC can help improve the development of the climate model of our country from the perspective of young talent calculation.

The demand in climate modeling is an important factor in promoting the development of HPC. As is mentioned in Nation in 2006, the first ocean-atmosphere coupling model developed by GFDL in 1969 marked a new age for scientific computing. In short, climate models can be considered as computer programs that apply a large number of physical rules, and is the mathematical expressions of different physics, chemistry and biology processes. The development of a climate model requires software engineers to develop efficient and flexible low-level code, generalized software platforms, and an open network protocol to support the output of network accessing. It also requires hardware engineers to maintain the supercomputing facilities to provide hardware support for simulation. However, it is nowadays a universal problem that the engineers for both scientific computing and software engineering are in great demand. According to the report from the American National Academy of Sciences (NAS), the number of Ph.D. degrees in atmospheric and ocean science has been around 600 annually in the last decade. However, the increased numbers for Ph.D. in computer science jumped from 800 in 2000 to 1700 in 2008, with only a very small part choosing to do the climate simulation, Therefore, NAS allocates in its National Strategy Report that more outstanding students should be cultivated into climate work, that the guidance to show the importance of climate modeling should be enhanced, and that various measures should be taken to absorb software engineers from other fields.

Even though that we have a very good foundation and situation in climate modeling, we have to admit the distance between us and developed countries in this area. Besides, we need to face the serious problems that the human sources in scientific computing and software engineering are in great demand. With the influences of climate change on human activities getting bigger, better climate models will be desired. As one of the most complicated tools of computer simulation ever in the history of human beings, climate modelling is an important method through which to understand and predict the rules of climate change, and is a matter of national issue that determines our goals in sustainable development as well as our rule in the international dialogue of climate change. Climate simulation and prediction have become a leading technique in supercomputing. The younger generations, especially those majoring in supercomputing, shall have the courage and braveness to devote their passion, wisdom, and energy to improve the status of our country in terms of the international competitions.

July 2015

Tianjun Zhou
Professor of the State Key Laboratory of Numerical
Modeling for Atmospheric Sciences
and Geophysical Fluid Dynamics (LASG)
Institute of Atmospheric Physics, Chinese Academy of Sciences
Member of the ASC Evaluation Committee

As an international platform for students to compete and communication, ASC has absorbed more and more students into the study and research on supercomputing technology. Through the good preparations for the competitions students can grasp much supercomputing knowledge and be ready to contribute to the development of the supercomputing. The test cases in ASC mainly focus on the essential real-world applications with a large potential for innovation, and can provide good opportunities for students to explore their potential and talent, and better understand HPC.

July 2015

Ming Li
Dean of the College of Mathematics
Taiyuan University of Technology
Chair of the ASC13 General Organization Committee

HPC has played more and more important roles in promoting technical innovation and scientific advance. Developing applications and cultivating talent are indispensable aspects that will be needed to transfer supercomputing hardware into industrial value. Through the organization of ASC, Inspur intends to let more talented young people understand and develop an interest in supercomputing at an early stage, promote the long-term and healthy development of supercomputing, and make supercomputers better serve scientific research and application.

June 2015

Jun Liu
General Manager, Inspur High Performance Computing
Chair of the ASC13 General Organization Committee

Preface

About ASC

So far, ASC is the world's largest annual student supercomputing competition. Based on the ASC competitions, this book aims to introduce knowledge of supercomputing techniques and helps explain rules and contents of the ASC competitions. This book has 14 chapters that are divided into three parts: Study, Competition, and Advances. 'Study' introduces the system, architecture, and evaluating approaches of supercomputers. 'Competitions' introduces the contents and rules of the ASC competition. 'Advances' contains information on how to analyze and optimize real scenario by referring to real cases.

As the first book that systematically introduces supercomputing competitions, this book is written by experts, and organized by the Asia Supercomputer Community and Inspur. The goal of the book is to foster interest in supercomputing amongst teenagers, and to provide technical support to researchers and engineers.

With the fast development of the information society, supercomputing has become one the most advanced technologies to meet the ever-increasing demand for computing capacity, and has been widely applied in many essential applications such as geophysics exploration, climate modeling, astronomy, and national defense. It is also an important method in problem solving in new areas from economy, government, enterprise, and online games. With the problems in essential issues getting bigger and more complicated, research starts to focus on developing Exaflops (EFLOPS) supercomputers.

In November 2015, Tianhe-2 supercomputer hit No.1 in the TOP 500 list, which is the sixth time in a row. With the breakthroughs in hardware technology, the peak performance of Tianhe-2 has been above 55 PFLOPS. However, the lack of professional and skilled talent is still a major bottleneck that prevents the further development of supercomputing. Application is the key factor that pushes forward the development of supercomputing, and it is the increasing demand for computing power from every industry that promotes the advancement of supercomputing technology. Due to the limitation of power consumption, we have to understand the application better to maximize its algorithmic parallelization, instead of simply relying on architectural parallelization. Therefore, talented researchers that can help understand the algorithm are in great demand. A lot of work is required to broadcast, cultivate and educate more talented young people.

For the purpose of broadcasting and developing supercomputing science and technology, especially in the universities to engage more talented young students, the supercomputing competitions were founded by ASC and Inspur. Initiated in China from 2012, the Asia Supercomputer (ASC) Community has received positive responses from students around the world, and gained wide support from supercomputing experts and institutions. In 2012, the first Chinese Supercomputing Competition was held. In 2013, the first Asia Supercomputer competition was held. And in 2014, ASC14 was held. Together with International Supercomputing Competition (ISC), and Supercomputing Committee (SC), ASC has become one of the most influential supercomputing competitions for university students around the world. Considering that ASC15 has engaged a lot of students from all over the world into the supercomputing area this book is expected to truly benefit and help advisors and students to understand better the competition, including the knowledge and rules.

The Book Structure

This book is born from the contributions of ASC, Inspur, as well as many distinguished specialists in supercomputing from all over the world.

The book contains three parts: Study, Competition and Advances. The Study part is an introductory part that can help readers understand the basic conceptions and methods of supercomputers, including the system architecture, network communication, development environment, and evaluation method. The Competition part introduces the ASC and its regulations, while the Advances part includes the sample questions of the ASC final.

The Study part starts with Chap. 1 that introduces the history of supercomputing, as well as its importance, architecture, and major challenge. Chapter 2 describes the principles of building a supercomputing system, including the related architecture and management. Chapter 3 focuses on the network environment while Chap. 4 introduces how to build the hardware, software, and development the environment, as well as the method used to set up the application environment. In Chap. 5, the evaluation systems for hardware, application performance, and projection method are explained.

Chapters 6 and 7 in the Competition part introduce the three major supercomputing competitions, namely the ASC, SC, and ISC, while Chap. 8 analyzes in detail the rules and regulations of ASC14.

Chapter 9 in the final part introduces methods for preparing for the preliminary stage. The remaining five chapters from Chaps. 10–14 are real examples that explain in detail how to optimize and boost the performance of given cases and how to analyze and use MIC platforms.

We believe that this book can be of great help to students and competition teams in particular so that their participants are more acquainted with the ASC competition.

Beijing, China Asia Supercomputer (ASC) Community

Acknowledgements

We would like to express our great gratitude to all the people that contributed and put their efforts into making this book happen, including Guoxing Yuan, Jifeng Yao, Xinhua Lin, and Tong Liu devoting themselves to completion of chapters. All members from the editing group worked hard together to collect and summarize their experiences and knowledge in order to complete the book.

The editing group of this book includes Wenjing Lv, Xinhua Lin, Weiwei Wang, Tong Liu, Guoxing Yuan, Jifeng Yao, Wong Foo Lam, Li Shen, Jerry Chou, Xiaofeng Liu, Minhua Wen, Jie He, Lian Jin, Yu Liu, Bowen Chen, Qing Zhang, Bo Shen, Zehuan Wang, Jianfeng Zhang, and Qing Ji.

Asia Supercomputer (ASC) Community

Contents

Part I
Study

Chapter 1
Development and Application of Supercomputing

1.1 Why Do We Study Supercomputing?

A supercomputer is a computer system that provides significantly higher computing power than normal personal computers. To provide a high computing performance, supercomputers are generally in the form of highly parallel systems. While supercomputers were mainly used by national labs and national defense agencies when they first came out, more and more industries are starting to use supercomputers for their research and development.

The increasing demand of computing power of different applications promotes the continuous development of supercomputers, and the requirements of different applications are quite distinct. For instance, numerical atmospheric simulation requires tight timelines; nuclear simulation requires high accuracy; intensive transaction processing requires high throughput; and some other applications might have demands on all three aspects of speed, accuracy, and throughput. These different challenges from different applications have been a major driving force for the development of supercomputing technologies.

In most cases, the demand for computing power in supercomputing applications is continuing to grow, and in general beyond the development of current supercomputer systems. Typical examples include numerical simulations in science and engineering. For these simulation problems we sometimes need to repeat the computation for a large number of times to get useful results. Moreover, users expect all of these simulation steps to be finished within an affordable timespan.

For example, in the manufacturing industry, a simulation should be finished within a few seconds or minutes if possible. During the design process, a simulation cycle that takes weeks is not acceptable. Only with an acceptable simulation time cycle, can the designers work in an efficient way. Moreover, when the simulation system becomes more complicated more time is needed to finish the simulation. Some applications have specific requirements on the time to solution. The most typical example is numerical weather prediction (NWP). It is meaningless if we

ASC Community, *The Student Supercomputer Challenge Guide*,
https://doi.org/10.1007/978-981-10-3731-3_1

spend two days completing one day's forecast, no matter how accurate it is. Therefore, the problems where we are not able to compute solutions within affordable time become big challenges.

Again, we can take NWP as an example. NWP is a widely used application that requires supercomputers. The modeling of the atmosphere is generally performed by discretizing the atmosphere into 3D grids. We then apply complex physics equations to advance in time. The variables (temperature, pressure, humidity, wind, etc.) within each grid are generally computed based on the values of the previous time step. To perform a meaningful weather simulation, we need to perform the computation within each grid for a large number of time steps, while the same computation also needs to be performed for all the grids in the 3D model. These two aspects add up to a huge quantity of computation that we need to perform.

To provide the weather forecast for several days, we need to have a large modeling domain, as the atmosphere could be affected by long-distance events. Assume we divide the entire atmosphere into grids in the size of "1 mile × 1 mile × 1 mile", and we simulate a layer of 10 miles (10 girds at the vertical axis), we need around 5 × 108 units to simulate the entire area. Assume that we need to perform 200 floating-point operations for each grid at each time step, then we need to perform 1011 floating-point operations to finish the simulation of one time step. Assume that we want to simulate 7 days using the time step of 1 min, we need to simulate for 104 time steps, which amounts to 1015 floating-point operations. For a 1 Gflops (giga-flops, 109 floating-point operations per second) computer, such a simulation would take 105 s (over 10 days). If we need to complete this simulation within 5 min, a 3.4 Tflops (tera-flops, 1012 floating-point operations per second) supercomputer is needed. Moreover, if we want to finish the simulation within 5 min with a higher resolution of grids (0.1 mile × 0.1 mile × 0.1 mile), the total number of floating-point operations increases to 1019 (note that the decrease of grid size would usually lead to a smaller time step), and we would need to have a 34 Pflops (peta-flops, 1015 floating-point operations per second) supercomputer (as a reference, the peak performance of Tianhe-2 supercomputer is 33.86 Pflops).

Predicting the movements of objects in the aerospace is another challenging problem that requires a huge amount of computing power. The objects attract each other due to gravitation law. To provide a movement prediction of each object, we need to find out all of the forces that affect it, and make a vector addition to get the composition. For a system with N objects, each object takes $N - 1$ forces, adding up to N^2 computations roughly. After determining the forces and the new locations of objects, we need to repeat the computation for the next time step. Take the galaxy as an example. There are around 10^{11} stars in total. Therefore, we need to perform 10^{22} computations for each time step. Even if we use a highly efficient approximation algorithm with the complexity of $N\log(N)$, the total amount of computation is still enormous (around 10^{12}), which would take an extremely long time to finish on one computer node. Assume that each computation takes only one micro second (this is an optimistic assumption, as each computation involves a large amount of addition and multiplication operations), when using the N^2 algorithm, one iteration would take 10^9 years to finish. When using the $N\log(N)$ algorithm, it can still take one year to finish one iteration.

The two examples we mentioned above (weather forecast and sky simulation) are the applications in the traditional scientific computing domain. People will continue to come up with more complex applications (such as virtual reality) that demand more and more computing power. In conclusion, no matter how fast the supercomputer is, there will always be new applications that require more.

In 1992, the US government initiated a 5-year plan of high performance computing and telecommunication. In that plan, for the first time, they proposed 12 urgent and challenging problems to solve, including fluid simulation, atmospheric modeling, fluid turbulence, pollution dispersion, human genome, ocean circulation, drug design, dyeing quantum dynamics, semiconductor modeling, supercomputer modeling, combustion systems, vision, and cognition. For some of problems, such as the human genome plan, we have already made meaningful progress; while for some other problems, we are still putting more efforts. In the twenty-first century, the following problems are still considered the most challenging: climate modeling and prediction; design improvements for automobiles, airplanes, and ships (related to fluid dynamics, fuel consumption, structural design and collision avoidance, etc.); bioinformatics; health and safety of the society; earthquake prediction; geophysics exploration; astrophysics (simulation of celestial evolution); materials science and nanotechnology; organization research (behavior simulation and counter-terrorism); nuclear simulation; numerical wind tunnel; and pollution detection and prevention.

In a supercomputer system, employing multiple processors to solve one problem is one common way to improve the computing speed, which has been studied for years. In such a method, the solving process is divided into several parts that could be executed in parallel, and each processor takes charge of one part. This computing method is called parallel computing. The computing platform, which is a parallel machine, could be a computer node that contains multiple customized processors, or a system that consists of multiple independent computers connected together.

In general, we expect to achieve significant speedup when taking the approach of parallel computing. In an ideal case, we expect N times speedup when using N computer nodes. However, in practical cases, it is difficult to decompose a problem into totally independent sub-tasks. The decomposed components usually need to exchange data with each other. Therefore, the actual speedup we can achieve is highly dependent on the parallelism within the algorithm.

Besides accelerating the solving process, the parallel computing approach can also enable the solution for large problems or more accurate results within an acceptable time frame. For example, when simulating a lot of physics phenomena, we discretize the problem domain into 2D or 3D grids, When using parallel computing with multiple computer nodes, we can process more grids within the same amount of time, thus leading to a more accurate depiction of the problem. Another factor is that a multi-node system would provide a significantly larger memory space than a single node. Therefore, it is possible to fit much larger problems.

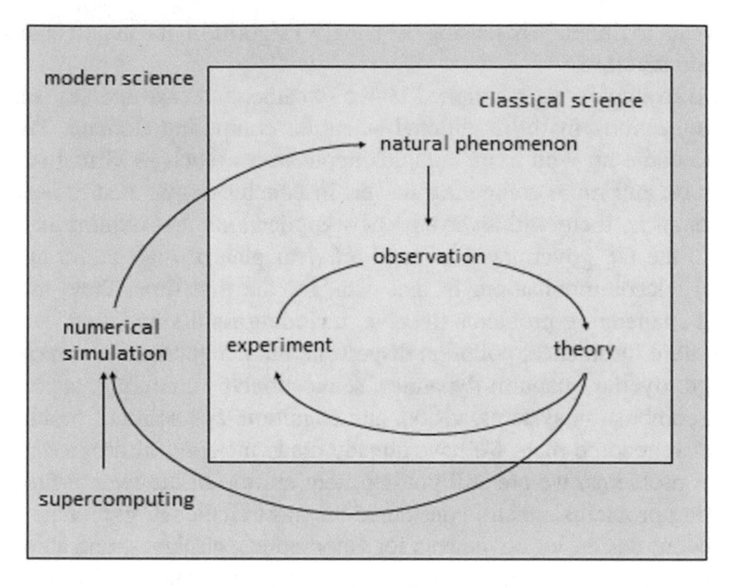

Fig. 1.1 Differences between classic science and contemporary (modern) science

After describing the urgent demands for supercomputers, now we will turn to the discussion about the main purposes of supercomputer research. In general, the main purposes of supercomputer research are as follows:

Firstly, computation has become the third method of exploring the world and making new discoveries, along with the scientific theory and experiments. Moreover, as time goes by, the computational approach is becoming more and more important.

The difference between classic science and contemporary science is shown in Fig. 1.1. Classic science is based on observation, theory, hypothesis and experimentation. In contrast, contemporary (modern) science adds numerical simulation to the process. By using supercomputers, we can use numerical simulation to verify our theory and hypothesis, instead of the traditional experiment approach. In many cases, the numerical simulation approach can be much more efficient.

Secondly, the development of supercomputers is a strong driver for the development of related computer science and technologies, especially for parallel architectures, high-bandwidth and large-capacity memory hierarchies, high-bandwidth and low-latency network, parallel programming methods, and so on.

Thirdly, research on supercomputers promotes the development of a number of inter-disciplinary subjects, such as computational mathematics, computational physics, bioinformatics, computational economics, computational electromagnetics, and computational geophysics.

Finally, the development of supercomputers is a good demonstration of the country's economic and technology power.

In order to reflect and promote the development of supercomputers in the world, from 1993, the TOP500 list[1] was announced twice a year, at the supercomputing conferences in Germany and the USA respectively. The TOP500 list includes the 500 fastest computer systems in the world, which are ranked by their LINPACK (the most widely accepted floating-point performance benchmark) performance. For a long time, the supercomputers in the USA dominated the top positions of the TOP500 list.

Coming into the twenty-first century, supercomputer technologies in Japan have developed quickly. The "Earth Simulator" supercomputer in Japan provided a peak performance of 35.6 Tflops, and had been the premier system in the world from 2003 to 2005. After that, the systems from the USA retook the top position. In 2011, the "King" supercomputer designed by NEC became the top system again with a peak performance of 8.1624 Pflops.

China's Tianhe-1A system, designed and built by the National University of Defense Technology (NUDT), ranked first on the TOP500 list in November 2010, with a peak performance of 4.7 Pflops. In June 2013, the Tianhe-2 system, designed by NUDT as the successor of Tianhe-1A, achieved a peak performance of 33.86 Pflops, and has been the fastest system in the world until today. In the TOP500 list of June 2014, there were 76 supercomputers in China, and was only superseded by the USA with 233 systems. In recent years, there has been a strong competition between different countries to develop the fastest supercomputers in the world. The TOP500 list is also changing rapidly.

Note that the TOP500 list takes a simple classification approach to dividing supercomputer systems into star-shape, vector machine, MPP (massively parallel processing), and cluster.

1.2 Development and Architecture Classification of Supercomputers

1.2.1 Development of Supercomputers

The term "supercomputer" was first used in the early 1970s. The first generation supercomputer had the architecture of a SIMD array of processors. In 1972, the USAs successfully developed the ILLIAC-IV processor array, consisting of an 8 × 8 array of 64 processors. The floating-point performance of this system is around 100 Mflops (mega-flops, 106 floating-point operations per second).

The second generation supercomputer was in the form of a pipelined vector machine. The Cray-1 system, which was designed by the Cray Company in 1976, is a typical example. The system includes 12 pipelined arithmetic units, each of which provides a different arithmetic function. The units can be divided into 4 groups, and

[1] https://www.top500.org/.

perform the computation in parallel. The peak performance is around 160 Mflops. Cray-1 was the first commercial supercomputer system on the market.

Depending on whether the operands and results are stored in registers or memories, vector machines can be divided into two types: the memory-to-memory type vector machine, and the register-to-register vector machine. Vector machines in the early days were mostly memory-to-memory type vector machines, such as ASC designed by the TI company in 1972, STAR-100 designed by CDC in 1973, and CYBER-205 as well as ETA-10 designed by CDC in 1980 and 1986 respectively. In 1976, the CRAY-1 vector machine adopted the register-to-register structure for the first time. Due to its excellent performance when processing short vectors and the simplified instruction system, the register-to-register architecture became the mainstream of vector machines.

Third generation supercomputers have shared-memory multi-processor systems that parallelize the computation in the MIMD (multiple instruction multiple data) way. The Cray-1 system in 1976 was a single-processor vector machine. Since then, to improve the performance of vector machines, people continued to increase the number of vector processors in the system. The Cray X-MP/2 in 1982 had 2 vector processors, and the Cray Y-MP system in 1984 had 4 vector processors. The number increased to 8 for the Cray Y-MP 816 system in 1988, and 16 for the C90 system afterwards. Coming into 1990s, the number of vector processors within one supercomputer system increased to a few hundreds. In the twenty-first century, the number of vector processors increased to the scale of a few thousands.

In order to balance the performance of vector processing and scalar processing, the current parallel vector processor systems mostly use high-performance superscalar processors. As an example, the "Earth Simulator" designed by NEC company in the beginning of the twenty-first century consists of 640 nodes, each of which contains 8 vector processors, 16 GB memory shared among the processors, a remote access control unit, and a I/O processor. The peak performance of the Earth Simulator's vector processor system was 40.96 Tflops, staying in the top position of the TOP500 list for around 3 years (2002–2005).

Fourth-generation supercomputers are MPP systems. The MPP system generally consists of tens of thousands of processors, and achieves a high performance through a high level of parallelism. We can build an MPP system using small nodes, NUMA, super nodes, or hybrid nodes with both vector and superscalar processors. In today's TOP500 list, MPP systems still take up a considerable portion.

Fifth-generation supercomputers are clusters, which are the most popular systems nowadays. The early clusters are homogeneous, while the current clusters are mostly heterogeneous, in the form of CPU + GPU or CPU + MIC (the Many Integrated Core architecture of Intel). The current number one system in the world, Tianhe-2 in China, provided a peak performance of 33.86 Pflops in the heterogeneous form of CPU + MIC.

1.2.2 Architecture Classification of Supercomputers

As supercomputers are highly parallel computer systems, we can apply the same classification of parallel computer architectures to supercomputers.

Depending on whether the parallel computing is achieved by SIMD (single instruction multiple data) or MIMD, and whether it uses shared or distributed memory, we can divide supercomputers into the following 4 types:

- shared-memory SIMD (SM-SIMD)
- distributed-memory SIMD (DM-SIMD)
- shared-memory MIMD (SM-MIMD)
- distributed-memory MIMD (DM-MIMD).

The early supercomputer systems perform parallel computing in the way of SIMD. As the processors are usually organized as an array, these systems are also called array processor. The memory in the array processor can be either shared-memory (SM-SIMD) or distributed memory (DM-SIMD). Array processor systems are usually special-purpose systems, and can only process one type of problems, but with high efficiency.

The single-processor vector machine has only one vector processor. However, the memory is shared among the vector processor, the scalar floating-point unit, and the scalar integer unit. Therefore, the single-processor vector machine is of the type SM-SIMD. These systems are general purpose, and provide a high efficiency when processing vector problems.

Currently supercomputer systems are mainly of the MIMD type. Multiple vector processor (MVP) systems contain MVPs that share the memory. Therefore, the MVP systems are of the SM-MIMD type. The symmetric multiprocessor (SMP) system is also of this type. Both MVP and SMP are considered as uniform memory access (UMA) systems, since the different processors in the system have the same access time to the memory.

The opposite of the UMA system is the NUMA (non-uniform memory access) system. The memory in a NUMA system is distributed, thus the memory access times for different processors depend on the location of the processor, and cannot be the same. Apparently, NUMA systems should be regarded as the DM-MIMD type. Note that the processors in a NUMA system could access remote memory by load-store instructions, so the system should have a unified logic memory space. Depending on whether the system provides hardware support for cache coherency, the NUMA systems can be further divided into CC-NUMA (cache-coherent) and NCC-NUMA (non-cache-coherent). If the memory part consists of only cache, we would have a COMA (cache only memory architecture) system.

If the processors in the system need to access remote memory by message-passing, then the system is called a NORMA (no remote memory access) system, which is also a DM-MIMD system. Different from NUMA systems, NORMA systems have multiple memory spaces, and each processor in the system is generally an independent computer node. The NORMA systems can be further divided into tightly coupled and loosely coupled systems. Clusters are generally loosely coupled NORMA systems, while MPP systems are usually tightly coupled NORMA systems.

MPP systems employ a large number of commodity computer nodes, and connect the different nodes using customized high-bandwidth and low-latency network. The memory is physically distributed, and need to use message passing to achieve inter-processor communication. MPP systems are tightly coupled parallel systems, and generally provide a good scalability. Typical examples of MPP include the Cray T3E and the IBM Blue Gene system.

In the cluster system, each node is an independent computer. The different nodes are connected using a commodity network. Each node runs a complete operating system, with an additional middleware to map different nodes into a unified pool of computing resources.

Figure 1.2 shows the structural classification of the supercomputer systems.

1.2.3 The Development Trend of Supercomputers

In the twenty-first century, with the increasing demand from various applications (weather/climate modeling, ocean modeling, nuclear simulation, vision, cognition, etc.), supercomputer systems are required to provide performance in the scale of Pflops or even Eflops (eta-flops, 10^{18} floating-point operations per second). Nowadays, we have quite a few Pflops supercomputers, mostly in the form of CPU + GPU or CPU + MIC heterogeneous clusters. The development of an Eflops supercomputer system is still in progress, with intensive competition between different countries.

Building an Eflops system is quite a challenge. The major reason is that, to achieve a performance in the scale of Eflops, we need to use a significantly larger number of processors and accelerators, which will lead to a tremendous increase in power consumption. Therefore, the development of processors with low-power and high-performance has become a key in the process of making the breakthrough. In addition to that, high-bandwidth and low-latency network, and heterogeneous architectures are also essential techniques to for building an Eflops-scale supercomputer.

We should also note that the development of E-scale supercomputers has to be supported by the corresponding developments in compilers and parallel programming languages. Programming tools are extremely important for improving the efficiency of the system and enabling the users to make use of E-scale computing resources.

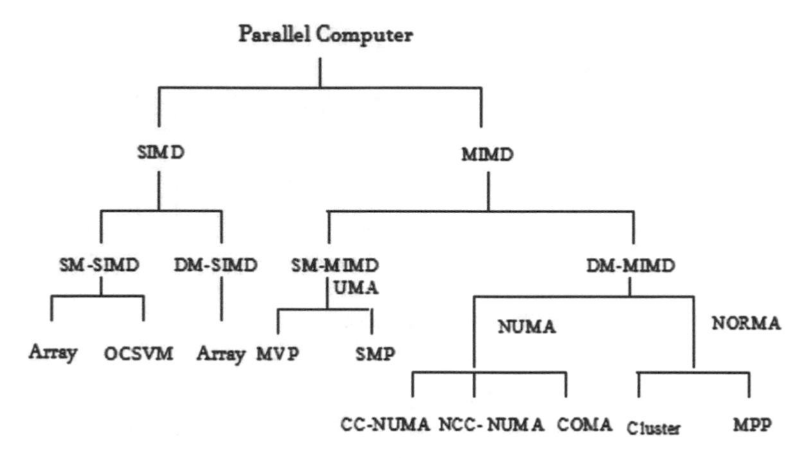

Fig. 1.2 Structural classification of the supercomputer system

1.2.4 The Rise and Development of Heterogeneous Architectures

Heterogeneous architectures date back to the 1980s. In many cases, the heterogeneous architecture enables the best match between the parallelism in the algorithm and the parallel computing power of the machine.

A heterogeneous computer system usually consists of three components: (1) a group of heterogeneous computer nodes, such as vector machines, MIMD machines, clusters, graphics processors, etc.; (2) a high-speed network (either a customized or commodity network) that connects these processors; (3) corresponding supporting software for programming the heterogeneous computers.

The basic idea of heterogeneous computing is to, firstly analyze the parallelism in the algorithm, secondly decompose the algorithm into different sub-tasks with different parallelism features, and thirdly map the different sub-tasks to different computing resources within the heterogeneous system, so as to achieve the best match between the algorithm and architecture and to minimize solution time.

Figure 1.3(i) shows an example of a loosely coupled heterogeneous computing system with distributed memory. The system consists of a vector machine, an MIMD machine, an SIMD machine and a workstation. Assume that within the algorithm, the portions of vector, MIMD, SIMD, and SISD computations stand for 30, 36, 24, and 10% of the total computation respectively. Assume that when running the right type of sub-tasks, the vector machine, the MIMD machine, the SIMD machine, and the workstation would achieve a speedup of 30, 36, 24, and 10 times respectively. T_s stands for the execution time on a serial machine. Then the time for running the same job on the heterogeneous system would be $T_p = (0.3T_s/30) + (0.36T_s/36) + (0.24T_s/24) + (0.1T_s/10) + T_c$, where T_c is the time for communication. Assume that T_c equals to 0.02 T_s here, then T_p equals to 0.06 T_s, which means we could get 16.67 times speedup over the original scenario.

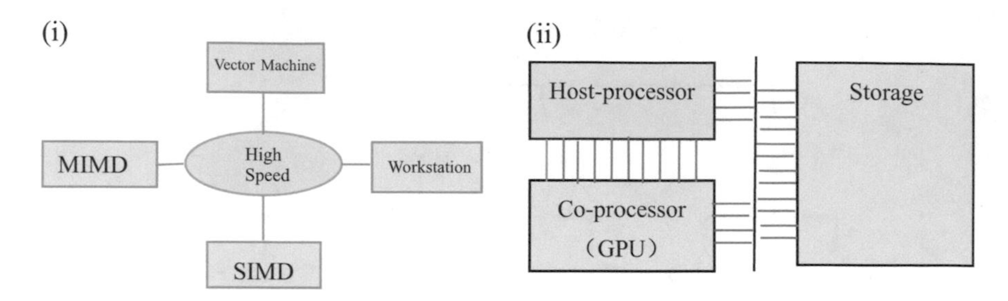

Fig. 1.3 Examples of tightly coupled and loosely coupled heterogeneous systems

Heterogeneous computing is widely used nowadays, in almost all the different scientific computing domains. Typical examples include image analysis, particle tracing, beam forming, and climate modeling. As all these different applications include different types of computation patterns, a heterogeneous system with different types of computing resources would be more suitable. We could take the image analysis system developed by the Hughes Institute and MIT as an example. The system contains three different layers, which corresponds to the requirements of the image analysis. The bottom layer is a SIMD bit stream network (4096-bit), which is used for pixel processing; the middle layer consists of 64 DSP chips, which perform pattern classification and other similar operations in the MIMD approach; the top layer is a general-purpose MIMD (coarse-grained) machine, which performs scenario, movement analysis, and other intelligent analysis functions.

The introduction of heterogeneous architectures into the supercomputer and the use of dedicated co-processors started in the twenty-first century. The major reason is that, since 1996, with the fast growing demands from applications, and the rapid development of VLSI and network technologies, the evolution of computer architectures was also accelerated. Although the superscalar techniques for exploiting instruction-level parallelism, dynamic prediction execution, and EPIC methods were successfully integrated into the commodity products, the design of the superscalar processors is becoming more and more complex, making it harder to further exploit instruction-level parallelism.

On the other hand, to provide a higher performance, the clock frequency of the processor chips becomes higher and higher, leading to the rise of power consumption and reduction of package density. Apparently, increasing the computing power by improving the processor's clock frequency and making use of instruction-level parallelism is not as efficient as before. Therefore, multi-threaded parallel computing and multi-core architectures become the new solutions. According to the practical results, multi-threading and data-level parallelism are becoming the key points for improving the parallel performance of the system. This change also demonstrates that parallel computing has become the mainstream technology of the current computer architectures.

As general-purpose processors are becoming more and more complicated and consuming more and more power, people start to think that it might be inefficient to continue pouring most resources on designing general-purpose processors. In contrast, special-purpose processors could be more powerful and energy-saving. Therefore, it could be quite helpful to offload more computing tasks to special-purpose coprocessors. The use of coprocessors is not a new technique. Since early days, microprocessors have been offloading a part of the job to floating-point coprocessors. The current term "coprocessor" indicates that these two kinds of processors are not symmetric. The coprocessors perform the computation for the master processors.

Using GPU for general-purpose computing is already a widely accepted method (this kind of GPU is usually called GPGPU). This is mainly because GPU has a high performance-price ratio. For example, compared with contemporary high-performance CPUs, the bandwidth and compute power of GPU could be one order of magnitude higher. In addition to that, the upgrade of GPU devices is also faster than CPU. The computing power is usually doubled every 6 months. As the programming tool of GPU is also improving quickly, GPU has a bright prospect as a general-purpose computing processor.

There are several reasons for the amazing performance of GPU. From the economic perspective, the performance demand of GPU is driven by the video games industry. From the technical perspective, GPU is designed for graphic processing, which is a kind of intensive computing application with a lot of data parallelism. Therefore, it would be much easier for GPU to improve floating-point performance by adding more processors. Moreover, as a coprocessor, GPU could largely ignore a large part of functions that the CPU needs to handle, and does not need to spend resources on branch prediction, and complex instruction scheduling units. The early GPU devices were only designed for supporting graphic processing, with a fixed hardware pipeline and support for single-precision floating-point operations only. In recent years, with new demands from game developers, the GPU hardware supports more and more different ways of programming. The result is that the GPU can now support general-purpose computing. The current GPGPU devices include units for double-precision floating-point and integer operations.

A modern GPU is like a massively parallel fine-grained multi-core chip with on-chip DRAM. For example, there are 8 streaming multiprocessors inside the NVIDIA GeForce 8800, and each streaming multiprocessor includes 16 SIMD cores.

Although the hardware architecture is of SIMD type, the programming model does not need to be exactly the same. The CUDA (Compute Unified Device Architecture) programming interface developed by NVIDIA is almost a subset of the C programming language. The remarkable advantage of GPU is that it could efficiently perform parallel floating-point computations. Figure 1.3(ii) shows the tightly coupled heterogeneous computing system with shared memory.

MIC (Many Integrated Core) is the latest parallel architecture developed by Intel Corporation, and is mainly applied in applications with a high level of parallelism in high-performance computing, workstation and data-center environments.

Compared with traditional processors, Intel MIC has smaller cores and hardware threads, and wider vector units, which could promote the entire performance more efficiently and meet the requirement for high parallelism. In addition, MIC provides a good compatibility, and supports programs of standard C, C++, and FORTRAN. Moreover, MIC and Intel Xeon CPUs share the same set of tools, compilers, and libraries. Since over 80% of supercomputers in the world use Intel Xeon CPUs, most of the developers can take advantage of their previous Xeon CPU experience.

In the TOP500 list of June 2014, there were 62 supercomputers that were heterogeneous systems with a certain kind of coprocessors. Among the 62 systems, 44 used NVIDIA GPU as a coprocessor, while 17 used Intel MIC (also known as Intel Xeon Phi) as the coprocessor.

1.3 Current Status and Challenges of Supercomputing Applications

The quality of supercomputing applications largely depends on the quality of parallel programming. However, parallel programming brings many more challenges than traditional serial programming. The major difficulty is that when doing parallel programming, programmers face a lot of different options that they do not need to consider in serial programming.

The programmers need to consider: which algorithm template to use; which parallel programming model to use; which parallel programming language to use; and what kind of supercomputer platform to use, etc. For current supercomputers, the parallelism is no longer hidden from the programmers. The software cannot achieve automatic performance boost along with the increase of CPU clock frequency. The programs would not have better performance if they were unable to use multiple cores within one processor.

1.3.1 Current Status of Supercomputing Applications

After more than 40 years of development, there are lots of improvements that have been achieved in parallel programming methods and techniques, detailed as follows.

1.3.1.1 Five Kinds of Parallel Programming Algorithm Templates Have Been Summarized

As mentioned earlier, parallel programming is a complex task. The programmer should firstly decide which kind of parallel algorithms should be used to solve the problem. Thankfully, although the history of the parallel algorithms is shorter than that of serial algorithms, there are still some well-defined templates for users.

(1) **The stage-based parallel algorithm**

In this algorithm, the program is divided into a number of stages. Each stage consists of the computing part, and the communication part. For the computing part, each process performs the computation independently; for the communication part, the processes exchange data by message passing. After that, the program goes to the next stage. Figure 1.4(i) shows the workflow of this algorithm.

The main advantage of the stage-based parallel algorithm is that it is easy for us to analyze the parallel performance of the program, due to the separation of computing and communication. However, the disadvantage is that we are not overlapping the computing part and the communication part, and it is generally difficult for to achieve good load balance among different processes.

There are two special cases of the stage-based parallel algorithm: synchronous iteration and asynchronous iteration. In synchronous iteration, different stages become a sequence of iterations inside one loop. As barrier synchronization is used, the stage $(j + 1)$ could not be started unless all the processes have finished stage j. However, in asynchronous iteration, since there is no barrier between different processes, the next stage for one process could be started as soon as this process finishes its current stage. We should notice that if there is any data dependency between processes, asynchronous iteration is not a suitable choice.

(2) **The divide-and-conquer algorithm**

The core concept of the divide-and-conquer algorithm is to continue to divide the problem into several sub-problems in a recursive way, until the sub-problem could be further decomposed. After the divide process, we assign these tasks into different nodes of a supercomputer to compute. After the computation is finished, we merge the results in a reverse pattern to the way that we divided the problem. Figure 1.4(ii) shows this process.

(3) **The macro-pipeline parallel algorithm**

This algorithm is based on function decomposition of the program. Firstly we divide the whole problem into several parts according to its feature, then we assign each part into one node. Communication in this algorithm is quite simple since data can only be passed between neighboring nodes in the pipeline. Figure 1.4(iii) shows the basic idea of this algorithm.

(4) **The master-slave parallel algorithm**

In this algorithm, the master process is responsible for the basic serial parts of the program, and for allocating tasks of the program into other slave processes. When one slave process completes its own work, the main process would be notified, and another task will be allocated to that process if necessary. In addition, the master also takes charge of summing up the sub-results. While the design of the master-slave algorithm is quite straightforward, the main process might become the

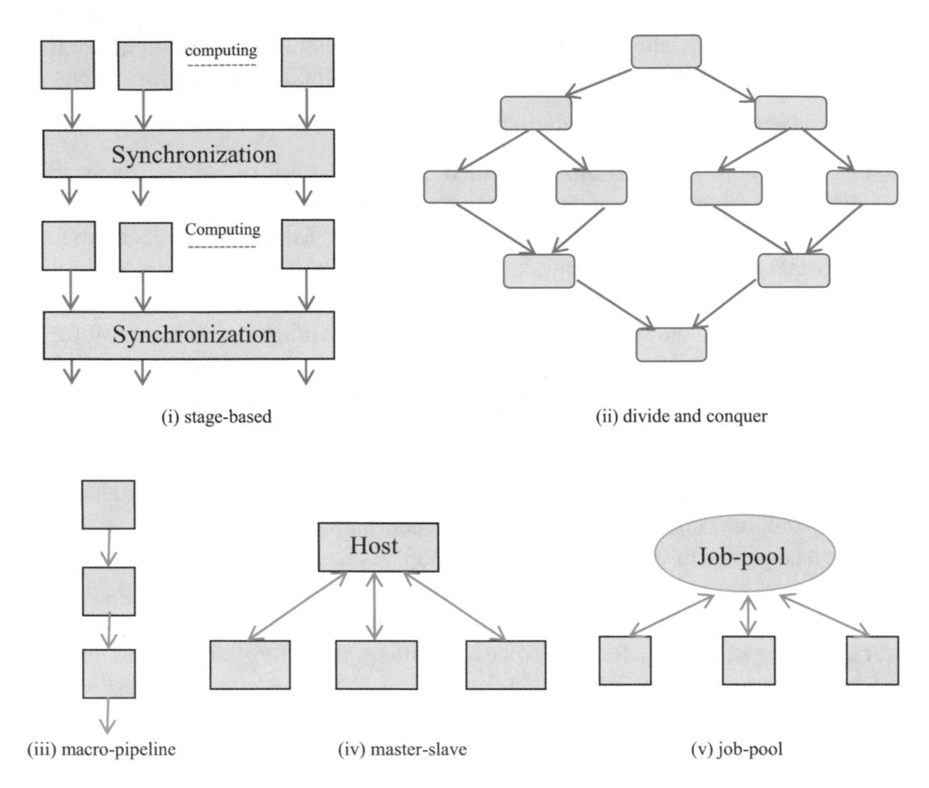

Fig. 1.4 Five kinds of parallel programming algorithm templates

bottleneck of the entire program. Note that in this algorithm, we generally take the static load balancing method, with the task allocation specified at the very beginning. Another feature of the algorithm is that the communication only exists between the master process and the slave processes. Figure 1.4(iv) shows the general workflow of this algorithm.

(5) **The job-pool parallel algorithm**

This algorithm is commonly used in the shared-variable model, and is generally implemented using global data structures. There might be just one job inside the pool at the beginning, and any of the idle processes that have been initiated can obtain a job from the pool and execute the job. During the execution of jobs, more jobs could be generated and put into the pool. When the pool becomes empty and processes are no longer generating new tasks, the execution of the program is finished. It is easy to achieve load balance in this algorithm, since the jobs are generated and allocated dynamically. The disadvantage is that it is difficult to achieve an efficient implementation for the processes to access the job pool concurrently. The job pool is usually implemented as an unordered set, a queue, or a priority queue. The algorithm is shown in Fig. 1.4(v).

1.3.1.2 Various Types of Parallel Programming Languages Are Available for Developers

There are various types of parallel programming languages available for developers which include:

(1) Multi-threading programming languages for the shared memory model, such as POSIX Threads, Java Threads, OpenMP.
(2) Message passing programming languages for the distributed memory model, such as MPI, PVM.
(3) Parallel programming languages for data-parallel designs, such as HPF (High Performance Fortran).
(4) Parallel programming languages for heterogeneous platforms, such as CUDA developed by NVIDIA.

1.3.1.3 Three Parallel Programming Models Are Available for Developers

There are three parallel programming models available for developers that include:

(1) The shared-variable model.
(2) The message-passing model.
(3) The data-parallel model.

In conclusion, with the rapid development of supercomputers as well as parallel programming techniques, supercomputers are widely used nowadays. In some application fields, such as the Human Genome Project, supercomputers have made significant progress in both scientific research and industry innovations, which enhances developers' confidence to apply supercomputers in even more domains.

1.3.2 Challenges of Supercomputing Applications

In supercomputing applications, the following problems are the key challenges the programmers should pay special attention to:

1.3.2.1 How to Write Highly Scalable and Portable Parallel Programs

The performance of a program is not only related to the application, but is also related to the hardware. It is hard to keep a linear speedup of the program when we increase the number of processors to a certain level. However, the number of cores available to programmers continues to be doubled every 18 months. Therefore, a successful software should be able to scale its performance with the increasing number of cores, in a similar way to the previous scaling of serial programs' performance with the increase of CPU clock frequency. Besides scalability, portability is another important feature of software that would be used long term, so that we can apply the same software on different supercomputers with reasonable performance. When writing a parallel program, we should take scalability and portability as the most important metrics from the very beginning of the design process. Following this principle, the resulting software would have a much longer life cycle.

1.3.2.2 How to Enable Automated Use of Effective Parallel Programming Techniques When Writing Parallel Programs

The automated use of parallel programming techniques during the programming process can largely improve the efficiency of the program. The current major parallel programming techniques that are proven to be effective include: incremental development; use of parallel data structures; efficient use of cache and other fast buffers in the system; extending the decomposition scheme from 1D to 2D or even 3D, so as to improve the parallelism and reduce communication and synchronization; and exploring different allocation methods, such as static/dynamic allocation, block allocation.

1.3.2.3 How to Enforce the Use of Parallel Programming Design Principles During the Programming Process

Why should we focus on the design principles? It is because, while the hardware architecture and the programming tools of parallel systems keep changing, the design principles will never be outdated.

The key principles that we should follow include the scalability and portability principle, the abstract model of parallel computing, static/dynamic allocation principles, and the owner-compute principle.

It is a challenging task for programmers to follow these principles when doing parallel programming.

1.3.2.4 How to Employ Suitable Optimization Techniques

Major optimization techniques include: an efficient utilization of the different components in the memory hierarchy, especially the fast buffers, such as cache; assembly-level code optimization.

1.3.2.5 How to Promote Interdisciplinary Collaboration

Supercomputing applications are usually related to other disciplines, besides computer science. Therefore, only with the close co-operation between the different disciplines, can we achieve the most efficient application software.

Finally we should mention that, with the development of supercomputers and corresponding applications, a new discipline called "Computational Science" is gradually maturing. With the interdisciplinary collaboration with other scientific domains, computational science and its related supercomputing technologies have been developing fast in recent years. The efforts on thesc areas would be a strong driving force for the improvement of technology competitiveness and innovation. The research and development of the supercomputer architectures are a key component in this process.

Chapter 2
Construction and Power Management of Supercomputing System

2.1 Components of a Supercomputing System

On the TOP500 list in November 2016, most HPC systems are either clusters (432 machines, 86.4% of the total) or MPP systems (68 machines, 13.6% of the total). The cluster architecture is now the most popular architecture among mainstream systems.

Usually, a typical HPC cluster system consists of five categories of computing (or network) devices and three types of network. The five categories of devices mainly refer to the login nodes, management nodes, compute nodes (including thin and fat nodes), switching equipment, I/O, and storage nodes. In addition, many current CPU/GPU/MIC heterogeneous HPC systems include the GPU/MIC accelerator nodes.

The login node serves as the gateway for users to access the cluster. Users typically log onto this node to compile and submit the job. The login node is the only entrance for external access to the powerful computing or storage capacity of the cluster, and the key point of the entire system. In order to guarantee the high availability of the login nodes, we should apply fault-tolerance approaches, such as hot backup dual-node technology, or at least RAID (Redundant Array of Independent Disks) technology to ensure the data security of the user nodes.

The login node generally does not require high computing power. In general, the entire cluster only needs to configure several rack-mounted servers as login nodes according to the requirements.

The management node is used to control all kinds of management strategies in the cluster. It is responsible for monitoring the robustness of each cluster node and the status of the interconnect network. Generally, the cluster management software also runs on this node.

As a result of the powerful processing capability of the modern servers, the job scheduler program (such as PBS) can also run on this node. Generally, the management node is also the key point of the computing network. If the management

ASC Community, *The Student Supercomputer Challenge Guide*,
https://doi.org/10.1007/978-981-10-3731-3_2

node fails, all the compute nodes would fail, and the submitted jobs would hang up. Therefore, the management node should also have hardware-redundant protections.

Similar to the login node, the management node does not require high computing power. In the entire cluster we generally only need to configure several rack-mounted servers (shown in Fig. 2.1) according to the requirements.

The compute node is the essential part of the cluster that performs the computations. The configuration of the compute node is generally decided by both the needs and budget. We can further classify the compute node into thin nodes and fat nodes.

Due to budget considerations, the main computing power of the cluster is provided by large numbers of thin nodes of the same configurations. Due to the large quantity of thin nodes, energy and space saving become some of the most important concerns of the customers. The blade server (shown in Fig. 2.2) has become the mainstream.

The fat node is usually used for the special applications that are difficult to perform domain decomposition or require particularly large memory space. The fat node (Fig. 2.3) generally uses more than 4 ways of processors, and large size of memory. Accordingly, the price of the fat node is also relatively higher.

The heterogeneous node, in recent years, it has already become a hot topic to adopt the heterogeneous computing system for general-purpose scientific and engineering computing. The current heterogeneous nodes usually adopt both CPU and GPU or MIC accelerators (Fig. 2.4). Making use of accelerators for parallel computing can significantly improve the computational efficiency in many cases.

The switching equipment, all the nodes in the cluster are connected through the network. Information and data are exchanged between the nodes via the switching equipment. In a large cluster, the switching equipment of the computing network often uses the large switches with hundreds of ports (Fig. 2.5).

The I/O node and storage device, to enable highly-efficient parallel processing, each compute node generally needs to access the data in parallel as well. Meanwhile, the concurrent computing processes generate large volumes of important data, and require professional storage devices that provide a large storage

Fig. 2.1 The rack-mounted server

Fig. 2.2 The blade server Inspur NX5440

Fig. 2.3 The 8-way fat server node TS850 from Inspur

space and security functions. The I/O node provides the interface between the storage devices and the compute nodes, to ensure the synchronization of data access.

The storage system does not only serve the role of data backup, but also provides the function of improving the read/write bandwidth during the computation process. In most software, we take the approach to keep the most frequently used data in memory, and to keep the less frequently used data on the disk storage. Even though the data stored on the hard disk is used less frequently than that stored in the memory, the amount of hard disk data is huge and needs to be updated at run time during the computation process. Therefore, we require a high read/write bandwidth.

Fig. 2.4 Inspur Yitian series heterogeneous supercomputing server NF5588

Fig. 2.5 The large all-gigabit-Ethernet switch

Generally, the local hard disk access speed cannot meet the requirements, and a specialized file storage system, as shown in Fig. 2.6, is needed.

The management network is used for the interconnection of the management nodes, compute nodes, and I/O nodes. Since the nodes connected by the management network are within the cluster, we do not require high bandwidth and low

Fig. 2.6 Inspur AS10000 mass storage system

latency, and we can tolerate a certain degree of over-subscription. Based on the above considerations, the gigabit Ethernet is a suitable candidate.

The computing network is used for the interconnection between the compute nodes and is specifically used for the inter-process communication (IPC) during the parallel computing process. One key component of parallel computing is the capability to exchange information with the other nodes in the cluster, which is often called IPCs. It requires a high-performance network for rapid exchange, thus requiring low latency and high bandwidth. The interconnect structure and bandwidth of the system are important factors that determine the system architecture, performance, and applicability to different applications. For example, mesh and multi-dimensional mesh structures are suitable for computational fluid dynamics (CFD) and other similar scientific computing applications. In these applications, the patterns in data partition and communication provide a good fit for the mesh structure, and can generally achieve a good efficiency of the system.

For large-scale computing tasks, parallel computing is the only solution to dramatically increase computing speed. With the development of CPU and memory technologies, the computing capability of a single node becomes more and more powerful. Therefore, in order to maximize the performance of each single node, we require high bandwidth and low latency of interconnect. The higher the bandwidth and the lower the latency the higher the computational efficiency that can be achieved from the entire system.

The computing network usually adopts the gigabit Ethernet, the IB networks or the 10-gigabit Ethernet.

The storage network provides the data access service for the nodes in the HPC cluster. In the HPC field, according to the different storage and access mode, we have a number of different storage models. At the lowest level, there are two ways to access the data: one is the file-level data access provided by the external file system, including NAS; the other is the data-block-level access, including DAS or SAN. SAN can either use IB storage or the Fibre Channel based on SCSI or SCSI RDMA (SRP) protocol.

2.2 Power Monitoring and Management of a Supercomputing System

With technology innovations, the server's initial cost continues to drop, while the power and other operation costs continues to rise. Therefore, improving the power efficiency of a computer system is a new focus for the development of supercomputers.

While the computing power of a supercomputer is significantly higher than normal computing systems, the energy consumption is even more astonishing. The power consumption of Tianhe-1 is 6000 kW, which costs up to 50 million CNY (Chinese Yuan) for the electricity every year. For a 50 peta-Flop supercomputer, the power consumption is estimated to be 20,000 kW, which is equivalent to 0.1% of the total electricity consumption of Shanghai. Facing the growing pressure on increasing energy consumption and the demand to be environmentally friendly, the design of supercomputers is increasingly focusing on the energy efficiency, that is, the computing capability per watt. To improve energy efficiency, we need to focus on two aspects: one is to design more energy-efficient supercomputer components; the other is to design a more energy-efficient cooling method.

Energy optimization is currently gaining extensive attention from the industry. Some people focus on the new system architectures, while most of us are still focusing on the refinement of the existing architectures.

Among the main components of the server, the processors consume a large portion of the entire power cost. Processor power consumption management methods include Dynamic Voltage Frequency Scaling (DVFS) and the dynamic sleep mechanism. With the maturing of low-power technology, almost all of the current mainstream operating systems have built-in power management strategies.

In most scientific computing applications, the power consumption of computation, communication, and synchronization are the major parts of the power consumption, with the I/O accounting for a small portion. Therefore, the processor, the power supply modules, the memory, and the fan become the major consumers of power. Figure 2.7 shows the power consumption ratio of different components when the server is in idle mode. Power consumption can be different for different servers and different workloads. However, in most cases, the power consumption of processors still accounts for the largest portion.

Currently there are a lot of power-reduction techniques, and a variety of effective power-management schemes have been proposed based on these techniques. These management schemes could be divided into two categories according to their mechanisms.

(a) Dynamic Resource Sleeping (DRS). The key idea of this technique is to shut down idle resources, such as components, devices or nodes to save energy, and awake them when needed. Currently most mainstream processors provide support for the DRS technology. The detailed standards of Resource Sleeping are described in the Advanced Configuration and Power Interface (ACPI). In addition to that, some memory devices also support dynamic shutdown. The PCI power management standards also describe the general ways to dynamically shutdown PIC devices. In a supercomputer system we could also put some idle nodes to the sleeping mode.
(b) Dynamic Speed Scaling. The nature of this technique is to dynamically adjust the executing rate of devices. There is a large amount of communication and synchronization in parallel computing. Therefore, in many cases, the fast components would have to wait for the slow components. In such scenarios, it

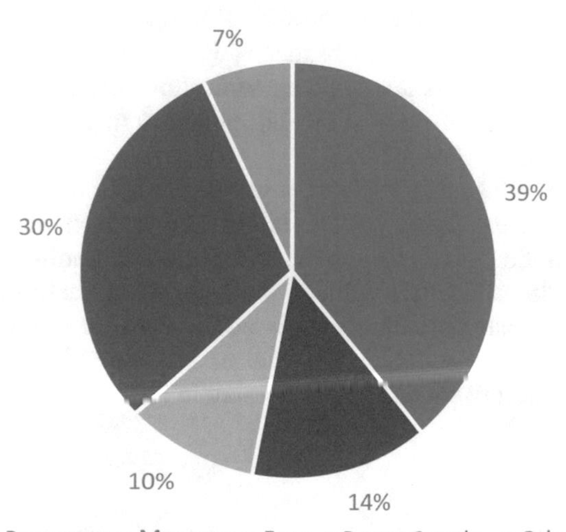

Fig. 2.7 Power consumption of different parts of a server in idle mode

is not necessary to keep the high rate of the fast components. By reducing the rate of fast components, we can reduce the power consumption while providing the same performance. The DVFS of processors is a typical DSS approach. Some memory and disk devices also support dynamic adjustment of the running rate.

Almost all current commercial processors support DSS and DRS as mentioned above.

In real supercomputing systems, processors will not always be busy processing computation tasks. It is would be a waste of energy to keep it running when there is no computation happening.

Therefore, the DRS techniques are introduced. Different processors support different sleeping modes with different power consumptions. In general, it takes more time and energy to wake up from a sleeping mode that consumes less energy. For instance, the Intel Xeon CPUs support the Enhanced Halt State technology. Besides the normal running state C0, the processor could also be in the sleeping mode C1, or even the enhanced deep sleeping mode technique. According to the idling record of the processor, the system would decide whether to put the processor into sleeping mode and will choose the right sleeping mode accordingly.

According to the experimental results, for test cases, using the C1E sleeping mode would take less energy and longer time to finish the job than using the C1 sleeping mode. However, for compute-intensive applications, the longer waking-up time of the C1E mode affects the efficiency more seriously. In such cases, while the C1E mode takes more time to compute, the energy consumption would also be more.

The optimization results by adjustment of the frequencies of the processors largely depend on the characteristics of the application. For communication-intensive applications, underclocking could bring about a 10% reduction in power consumption. However, for compute-intensive applications, underclocking might even lead to increased power consumption.

Therefore, we should make adjustments of the processor frequencies according to the computation and communication patterns of the parallel program. In supercomputing, there is still a big potential in power optimization by dynamic speed scaling of processors.

There are two levels of power consumption monitoring for clusters. The first is at the cabinet level. A PDU is installed in each cabinet to measure the power consumption within the cabinet. The monitoring software can then collect the data from all the cabinets to achieve a picture of the entire system. This also provides the direct data for estimating the power consumption cost of the system.

The second level is at the node level. We could either employ professional monitoring software or some simple tools in the OS to perform the monitoring of the node.

2.3 Building a Performance-Balanced HPC System

2.3.1 A Performance-Balanced HPC System

According to Amdahl's law, when the CPU performance increases by 10 times with the I/O performance remaining the same, the overall performance of the system can only increase fivefold; similarly when the CPU performance increases by 100 times without improving the I/O performance, the overall performance of the system can only increase tenfold. Therefore, the balance among the various components of the system is extremely important. If the poor performance of some components becomes a bottleneck of the system, the overall performance will reduce, and we will not be able to take full advantage of the other components in the system.

Generally speaking, there are two aspects about design balance of a supercomputer system. One is the architectural balance among the capabilities of different hardware components in the system. The other is the software level balance, which mainly refers to the load balance of different nodes achieved through effective management of different resources.

The supercomputer mainly consists of the computation components, the storage components, and the interconnect components. To achieve a balanced system requires coordination of the performance of these three components, so as to avoid both redundancy and bottleneck under the specific workload.

To build a well-balanced HPC system, we need to take the following aspects into consideration.

2.3.1.1 The Intra-node Configuration

(1) *The homogeneous CPU type*. We should keep the intra-node configuration as consistent as possible in order to improve the data exchange capabilities and the efficiency of resource utilization. For example, if the node is 2-way, we should keep the same memory configuration for both processors, so as to avoid the memory difference, the performance difference, and resulting reduction of the processing capability of the node. Specifically, for a 2-way node, deploying 24G memory for both processors will bring better efficiency than deploying 24G memory for 1 processor and a 12G memory for the other.

(2) *The heterogeneous type*. In a heterogeneous computing system using both the CPU processors and GPU accelerators, the host memory should be in general larger than or at least equal to the GPU onboard memory. One good idea is to build a unified and shared memory space between the CPU and the GPU so as to achieve seamless interaction between the two. We should keep a balanced ratio between the number of CPU cores and the number of GPU cards, as well as a good ratio between the GPU onboard memory bandwidth and the host memory bandwidth.

2.3.1.2 The Inter-node Configuration

For the compute nodes of the same category, we should keep a same balanced configuration for major hardware parameters, such as the number of processors, the number of cores, memory size, and the architecture of the node (homogeneous or heterogeneous).

2.3.1.3 The Network

Similarly, connections between the same types of nodes should adopt the same kind of interconnect, such as gigabit Ethernet, 10-gigabit Ethernet, IB networks.

For switches, we should either use full-bandwidth cascading or half-bandwidth cascading for the entire system. Otherwise, the different bandwidth in different connections might lead to communication congestion.

2.3.1.4 The Configuration of Nodes Executing the Same Task

For nodes that execute the same task, we should keep their configuration consistent to achieve a faster speed when processing and exchanging the data. If the hardware consistency cannot be achieved due to practical reasons, we can use cluster management software to allocate the job to the compute nodes with the same configuration, so as to avoid performance bottleneck caused by certain nodes within the cluster.

2.3.1.5 The Balance Among Different Devices

We should also consider the balance between the compute nodes, the I/O nodes, and the storage devices. The storage system within a HPC system is responsible for not only the data backup but also the increase of the read/write bandwidth during the computing process.

For applications that mainly perform computations, and seldom store or read data from disk, it is unnecessary to deploy a high-end I/O node. In contrast, for applications that involve large-volume data read and write, such as the genetic and geophysics exploration applications, we need to deploy a powerful I/O node to satisfy the high I/O demand.

2.3.1.6 The Power Consumption

Based on the practical situation of the computer room, when designing the system layout, we need to distribute the devices in a balanced way, so as to avoid the power consumption hot-spots in the room.

Under the constraint of satisfying all function requirements of the system, we should combine devices with high-energy consumption and devices with low-energy consumption, and combine the high-density devices with the low-density devices. For example, putting the blade servers and rack servers into the same cabinet can avoid huge power consumption differences among different locations and the possible local overheat or local waste of power resource.

In addition, to make efficient utilization of resources, we could place the high-density equipment with high-power consumptions close to the cooling equipment in the computer room.

2.3.2 *How to Manage the HPC System*

HPC systems have been widely used in many different domains. Meanwhile, its management has also become an important issue. With the continuous improvement of the computing performance, the scale of the compute node has also kept on expanding. A typical HPC system nowadays can have thousands of compute nodes, different kinds of computing resources (CPU/GPU/MIC), a complicated configuration of management software, hundreds of users and jobs running in parallel. All these quantitative increases would eventually lead to the qualitative change on the challenge of management for HPC systems.

In order to achieve higher efficiency, an HPC system requires an integrated management of the computing resources, network resources, storage resources, energy resources, and job executions, as shown in Fig. 2.8. For the computing resources, the heterogeneous accelerators and the multi-core architectures of the processors have increased the complexity of the management. For the network resources, due to the increasing scale of the interconnection, the performance impact of the mapping from the interconnect topology to the parallel program

Delivery & Service	Installation	Tuning	Training
Application Development	CPU parallel application	GPU parallel application	MIC parallel application
Application Tuning	Life science	Oil & Gas	Weather forecasting
	Computational chemistry	CAD/CAE	Deep learning
System Management	Monitoring & Management	Job scheduling	OS deployment
	KVM	Load balance	HA
Hardware	Servers	Storage	Network
Infrastructure	Cabinet	Power & Cooling	Datacenter monitoring

Fig. 2.8 The overall structure of an HPC system

becomes an important factor. For the storage resources, the increase of the different levels within the system, the increase of levels of cache within the multi-core processor, and the distributed structure of the file system all bring tough challenges for storage management. For job scheduling, the increased types of parallel programming models require more effective management of the jobs of multiple models.

In order to achieve high availability of the HPC systems, the management needs to include flexible extensions, which can support a quick adjustment according to the change on system management objectives and application objectives. Meanwhile, the growing demand from the parallel applications brings a heavier and more complex workload to the large-scale processing system. Moreover, applications from different domains generally demonstrate different parallel features in both time and space. All these different challenges result in a significantly increased complexity of job scheduling, making it difficult to scale the throughput with the scale of the system. Therefore, we need to develop a scheduling approach that is more suitable for the large-scale parallel processing system features and the application characteristics, and to enhance the adaptability of the scheduling policies, so as to achieve a high efficiency.

As for China's local HPC manufacturers, they need to focus more on the innovations in the aspects of the applications and product technologies, and apply the HPC systems into the practical industries, such as medicine, chemistry, materials, physics, CAE, oil and gas, weather forecasting, environment protection, academics, and scientific research. In addition, HPC manufacturers also need to develop a fair amount of management software (e.g. cluster monitoring software, cluster job scheduling software, system deployment software, and system backup software) so as to make the operation more convenient for the users, and to improve the practical efficiency of the system.

2.3.3 Monitoring Management

The cluster monitoring software mainly achieves a 1-to-1 mapping of the entire system, which provides a unified platform for the system administrator to monitor each single node in the cluster. Through the software, the administrator can choose to monitor the nodes as groups, or to monitor 1 specific node. The current mainstream cluster monitoring software generally adopts the B/S structure to achieve highly efficient remote management, and provide functions such as the group management, parallel operation, 1-to-1 mapping, and quasi-3D graphical interface. Some monitoring software even provide functions such as accounting, and raising alarms, thus leading to a significant decrease in users' management costs and improvement of management efficiency.

Through the monitoring of various components and patterns of the cluster system, such as the status of processor, memory, network card, workload, memory page, process context, swap partition, disk capacity, and disk I/O, the monitoring software can collect real-time data to help the cluster administrators obtain the basic data, and to analyze and manage the system.

The cluster monitoring software provides an integrated interface for the administrator to perform centralized management of the cluster system and the related resources. For example, the administrator can create a user or user group, and turn off or restart a cluster node through the software interface, rather than operating on the physical machine. The monitoring log provided by the software also helps the administrator to locate and resolve problems in a better way.

A typical HPC cluster can be very complex in structure, and generally consists of a large number of different nodes and devices. Therefore, as shown in Fig. 2.9, it is impractical to rely on the administrators to manually monitor and fix the problems of the system. Instead, we can use an alarm system, which can actively scan the cluster system for failures and automatically report the problems to the administrators through light, sound, email, or text messages. In this way, the administrator of the system can quickly identify the problematic node and perform the maintenance work. The alarm system is a smart combination of the personnel, the technology, and the equipment, and is regarded as the most effective measures so far. Users can get high reliability of the system with a relative low cost.

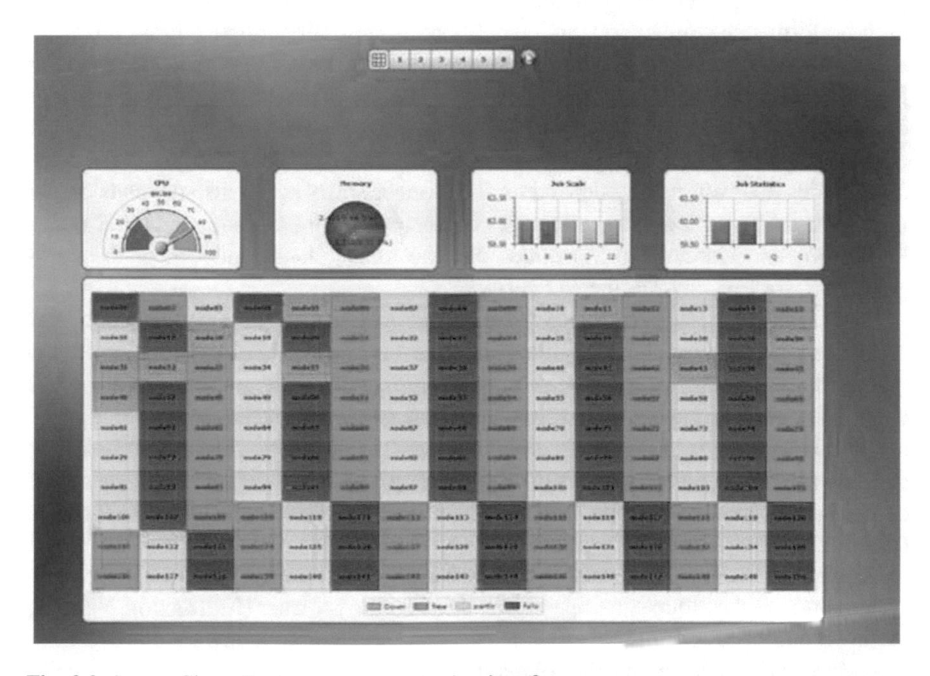

Fig. 2.9 Inspur ClusterEngine cluster monitoring interface

2.3.4 Job Scheduling

The cluster platform is widely applied in a variety of application domains, including the basic science research (such as computational chemistry, geophysics, artificial intelligence), public service (such as numerical weather forecasting, earthquake prediction), industrial and engineering computing (such as aerospace, automobile design, geological exploration, etc.), and data processing (such as finance, e-government). Therefore, the job scheduling system must be able to support various applications with different characteristics, ranging from engineering computing applications to basic scientific research computations. Based on the above considerations, the job scheduling system needs to provide proper interface and background support for both commercial software and open-source software, so as to ensure the stability of system when running different applications.

The job scheduling system is the channel for the users to access the HPC platform. We use the login node as the only interface to access the computing service, and also a barrier to isolate the users from the back-end system to improve reliability and manageability. At the management network layer, the job scheduling system provides a communication platform for the deployment, monitoring, scheduling, and management of the system. At the computing network layer, the job scheduling system provides the data communication between the high-performance application and the parallel-computing applications, so as to reduce the latency and to increase the bandwidth. At the storage network layer, the job scheduling system provides the high I/O throughput and high-bandwidth communication access for the storage servers and storage devices to ensure the high concurrency and large throughput of the storage usage. The storage access from other nodes does not merely depend on the storage network but also depends on the management network or computing network between the I/O nodes and other nodes (depending on whether the I/O nodes use the management network or the computing network to provide the I/O service).

The traditional job scheduling systems use commands to schedule jobs. The scheduling process can involve numerous and complicated combinations of commands therefore users need to be computer experts with advanced skills. The current job scheduling systems adopt a unified graphical interface to submit jobs, and the operation is more convenient and less error prone. The job scheduling systems provide concrete support for the running and scheduling of large-scale and complicated computation jobs, improve the efficiency of the cluster, and reduce the operation cost significantly.

Nowadays, the mainstream job scheduling software usually uses the B/S architecture, and conducts the operations through the browser (IE, Maxton, Firefox, etc.). Through the scheduling software, we can manage the hardware and software resources of the system and the jobs submitted by users. According to the cluster resource usage, we can then schedule the jobs to improve the resource utilization rate and the computation efficiency.

2.3.5 *The 10-Teraflops HPC System Design*

A computation capability of 10 teraflops (Tflops) is generally considered as the starting point for a modern HPC system. A cluster with a performance of over 10 Tflops already demonstrates the basic features of an HPC system. On the other hand, as 10 Tflops is only the starting point, the structure of the system does not need to be as complicated as a 100-Tflops system or a system with even higher performance. For example, the management node and login node do not need to be physically separate. Instead, they can share the same physical node with the compute node. In addition, the requirements on network and uninterruptible power supply are also relaxed. In general, we can build a 10-Tflops system in two ways: the CPU homogeneous approach and the heterogeneous approach.

2.3.5.1 Build the 10-Tflops HPC System

When building a 10-Tflops HPC system, we should mainly consider the following aspects:

Computing Power
As the main part of the entire cluster, the compute nodes are the basis of the entire system. The performance of the compute nodes directly determines the overall performance. Therefore, when choosing the processors, we should select the processors with more computing cores within the same chip size, and the processors with higher performance and less power consumption, such as the 22-nm Intel Ivy-bridge processors.
The compute node is a complete computer system that involves the interaction among all various components. In high-performance computing, almost all the computing software relies heavily on the frequent interaction between the CPU and the memory. The high performance of the CPU can only be made full use of when matched with the suitable memory. Therefore, when building the system, we should select the large-capacity and high-bandwidth memory to improve the performance and provide the fast memory access channel for each CPU.
Management
In terms of management, the HPC system handles different hardware devices, complex software configurations, and an increasing number of users and jobs. A good cluster monitoring and management system for the system can help to improve the management efficiency. The management system should include functions for the management of each node, the management of operating system and software, and the management of the cooling system and the room environment, for example.
The Network
The current mainstream of the HPC network includes the IB (InfiniBand) network and the GbE (Gigabit Ethernet) network.

The InfiniBand is a unified interconnection structure, and can be used for the storage I/O, the network I/O, and the IPC. Due to its high reliability, availability, scalability and the high performance, the InfiniBand can be used to interconnect the disk array, SANs, LANs, servers and the cluster servers, and can also be used to connect to the external network (such as WAN, VPN, Internet). Since the InfiniBand can provide the high-bandwidth and low-latency transmission within a relatively short distance and support the redundant I/O channels in a single or multiple interconnected networks, the IB network can still keep the data center running when having partial failures. In recent years, the IB network has become more and more popular among the HPC systems. In the June 2013 TOP500 ranking list, the number of machines using the IB network reached 205, accounting for 41% of the total.

The GbE network is defined by the IEEE 802.3-2005 standard. The GbE network used to be the most popular interconnect in the HPC systems. However, with the rapid development of the IB network, the market share of GbE has decreased slightly. In the 2013 TOP500 list, 140 machines adopt the GbE network, accounting for 28% of the total.

The Storage

The HPC storage system not only plays the role of data backupbut can also improve the read/write bandwidth during the computation process. The current mainstream storage systems include DAS (Direct-Attached Storage), NAS (Network Attached Storage), SAN (Storage Area Networks), and parallel file system storage.

In a DAS storage modethe RAID disk array is mounted directly onto the server network systems, the storage devices are directly connected to the server through the cable (usually the SCSI interface cable), and the I/O request is sent directly to the storage device. Based on the server, the DAS storage is a hardware stack without any storage operating system, thus needs the support from the corresponding server operating system.

In the NAS structurethe storage system is no longer attached to a particular server or client by the I/O bus, but is directly connected to the network through the network interface and accessed by the network. The NAS storage will store the files from different platforms onto 1 NAS device, enabling the network administrators to centrally manage large amounts of data and reducing the cost of maintenance.

SAN is a high-speed network or sub-networkproviding data transmission between the computer and the storage system. A SAN network consists of the communication structure responsible for network connections, the management layer responsible for organizing the connections, the storage components and the computer system, thus ensuring the reliability and effects of the data transmissions.

The parallel file storage system is mainly used to provide the high-bandwidth storage capacity management. The parallel file system is a network file system used in the multi-node environment. The data of a single file is distributed onto the different I/O nodes in a striped formand supports concurrent access from multiple processes on multiple machines. The parallel file system also supports the separated storage of the metadata and the raw data, and provides a single unified directory space.

The Operating System

As for the operating system, from the statistical results of the TOP500 list, more and more supercomputers are using the open-source Linux operating system. It is not only because of the low cost in software (Linux is an open-source operating system), training and porting, but also because Linux can provide a stable and expandable platform that allows users to continuously improve the system performance according to their own needs.

The Cluster Network Security

To facilitate the management of the cluster, the administrator usually connects the office network and the cluster management network. However, since the office network is usually interconnected with the Internet, this "fake physical isolation" connection is at a considerable risk of security. Despite of the convenience, the connection between the office network and the cluster management network also opens a gateway for the hackers to attack. The attacks on the cluster system mainly include application-level attacks, network-level attacks, and infrastructure-level attacks.

A variety of network threats from the outside can be prevented by installing the firewall and VPN servers in the front-end of the cluster system. The architecture with the protective functions is shown in Fig. 2.10.

(1) The firewall. The current firewall products on the market can be divided into two categories, the hardware firewall and the software firewall. The software firewall is mostly based on PC architecture, and may adopt an optimized OS as its operating platform. The features of the software firewall include good scalability, strong adaptability, easy to upgrade, and far less cost than the hardware-based firewall.

Most hardware firewalls are based on ASIC (Application-Specific Integrated Circuit). The advantages over the software firewall include faster speed, better

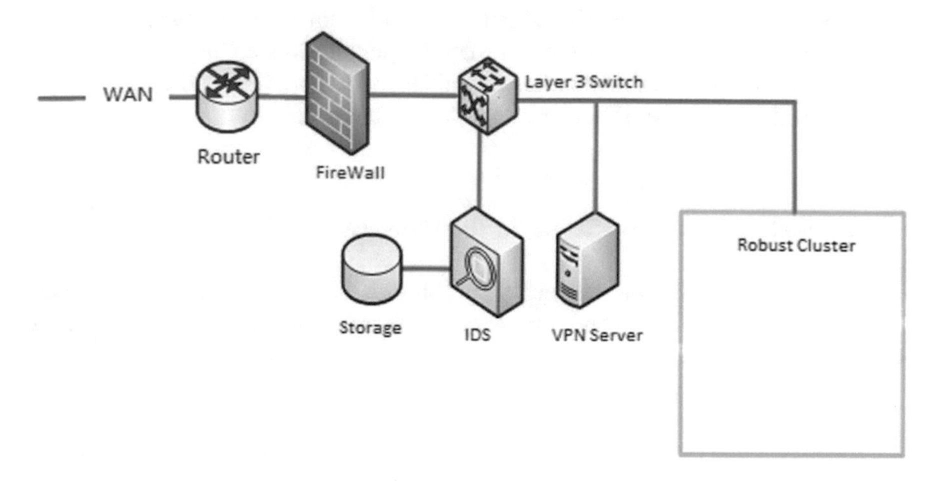

Fig. 2.10 The cluster system protection

stability, and higher safety factor. However, the hardware option is more expensive, less scalable, and more difficult to upgrade compared with the software firewall.

Due to its high cost, the hardware firewall is used less frequently in practical systems. We usually use the open-source software firewall for the servers. There are many open-source firewalls based on the Linux operating system, such as Zentyal, pfSense, IPFire, SmoothWall, ClearOS, and Iptables. Specifically, Iptables, which is embedded in the Linux kernel, is a powerful firewall that supports configuration of various rules. With a proper configuration of the IP rules, the system security can be significantly improved. In addition, we can also configure the Iptables to prevent IP spoofing and Dos attack problems.

(2) The VPN server. VPN (Virtual Private Network) is a remote access technology, which builds the private network based on the public network link. It is called a virtual network, mainly because in the entire VPN network, there is no end-to-end physical link used in the traditional private network. The VPN uses the logical network based on the network platforms (such as ATM, Frame Relay) provided by the public network service provider. The transmission of the user data is on the logical link. The VPN provides functions such as the encapsulation and encryption of the data, and authentication of the user identity across shared or public network, and is an extension of the traditional private network. OpenVPN is the most mature and most stable open-source VPN technology on Linux, and allows the establishment of VPN using the pre-set private keys, third-party certificates, or username/password for authentication. The OpenVPN uses the OpenSSL encryption library, and the SSLv3/TLSv1 agreement, and can support a number of different operating systems, such as Linux, xBSD, Mac OS X, and Windows 2000/XP.

The VPN uses technologies such as the tunnel, encryption/decryption, key management, and the user/device authentication. OpenVPN has a number of inherent security features. For example, it can be run in the user space without modifying the kernel and network protocol stack; it gives up the root privileges and runs in the chroot mode after the initialization; and it uses the mlockall to prevent the exchange of the sensitive data to the disk.

2.3.5.2 Build a 10-Tflops Cluster in the Homogeneous Mode

For a 10 Tflops or even more powerful CPU cluster, the heat dissipation, noise, and power supply are already beyond normal office environments, and need to be placed into specific computer room, which usually have state standards to follow.

The Compute Node

Taking the AVX-supported Intel SNB (Sandy-Bridge) platform into consideration, up to 40 2-way nodes are sufficient to provide the 10-Tflops computing performance. The detailed calculation is as follows:

The floating point capability per compute node equals:

2 (the CPU number per node)

x8 (flops per clock cycle/the calculated cores number)

x8 (the cores number per CPU).

If the frequency is 2.0 GHz (the starting frequency of the E5-EP processors on the general SNB platform), the computing power of a 2-way compute node is more than 256 Gflops, and 39 compute nodes can already provide the 10 Tflops (10 Tflops = 10 × 1000 Gflops) peak performance.

As for the type of the compute nodes, we have the option of the rack-mount servers or the blade servers. If adopting the rack-mounted server, we will need at least 39U rack space to meet the needs of the computing capability; assuming we are building our system using 1U standard rack-mounted 2-way servers, we need at least 1 standard 42U cabinet. However, such a design leaves no space between the nodes, and the heat dissipation could become a problem. Therefore, if we can afford more space, we usually use at least two cabinets, and reserve the space for the switches and storages. We can also adopt the high-density twin rack-mounted server. Generally, 4 2-way nodes could be accommodated in a 2U space and the density is doubled so that all the compute nodes can be housed in 1 cabinet. If we choose the blade servers, taking the current popular high-density blade server Inspur NX5440 as an example, an 8U space can accommodate 20 2-way nodes. Hence, two NX5440 servers can meet all the requirement of the computing power.

The Management Node

Generally, the number of users for a 10-Tflops cluster should be less than 20. We do not need to use the expensive commercial cluster software, as the challenge for the monitoring, management, and job scheduling would be relatively small. Instead, we can use the free and open-source software to meet the demands, such as openPBS. As mentioned above, due to the relatively small pressure on the hardware, the management node can share the same physical machine with the login node.

The Login Node

The login node and the management node can share the same machine, which can use a lower configuration than the compute node, such as a single rack-mounted server.

The Interconnect

When using the IB network, we need to consider different situations of the compute nodes. If adopting the rack-mounted nodes, we need to use a separate IB switch configuration. The number of the general 1U IB switch interfaces is configured to be 36. Over 36 interfaces will need to be configured with more switches. In this case, we need to consider whether the IB network needs the full line rate. If we need to do the full line rate, we also need to consider how to make the most economic interconnection among the servers and among the switches. If adopting the blade

nodes, we can directly interconnect the two blades through the IB network because the blade node generally has its own IB switch module.
When using the gigabit Ethernet as the computing network, we only need 1 48-port gigabit Ethernet switch. In the case of the using blade nodes (taking the NX5440 as an example), we can just interconnect the 10-gigabit uplink ports of the gigabit switch modules of 2 NX5440 blades.

The Management Network

We usually use the gigabit network as the management network, and the interconnection method is as mentioned above.

The Storage

For typical 10-Tflops HPC clusters, the I/O pressure is generally small, and we can take different approaches according to different situations. For example, if the output data is small, we can directly use the hard disk of the management node to do NFS data sharing and regularly back up the result data. On the other hand, if the output data is large, we recommend specialized storage devices, such as the storage servers or the disk arrays, and using the XFS file system.

2.3.5.3 Build a 10-Tflops Cluster in the Heterogeneous Mode

In contrast to the homogeneous mode, it is simpler to build a 10-Tflops cluster in the heterogeneous mode. For the current 1-Tflops accelerator cards, such as NVIDIA Tesla K20M or Intel Xeon Phi, 10 accelerator cards are enough to provide the required 10-Tflops computing power. If combined with the computing power of the CPUs, the number of needed nodes will be significantly smaller.

When adopting the heterogeneous mode to build the clusters, we need to consider the application features and performance balance, as well as what kind of hardware architecture to use. For example, if the application demonstrates a lot of thread-level parallelism (e.g. using OPENMP heavily) and scales better within a single node, we should consider using high-density heterogeneous servers, with two CPUs and four GPU cards in a single compute node.

If the application scales better across multiple nodes, and we hope to do large-scale CPU computation as well as GPU computation, we can consider equipping each node with 2 CPUs and 1 GPU.

Even if adding only 1 GPU accelerator card to each node, to meet the same 10-Tflops performance requirements, we only need 12 compute nodes.

When building the 10-Tflops heterogeneous HPC system, the considerations about the compute nodes and network are the same as the CPU homogeneous cluster. However, as for the management node and login node, they must include the support for GPU management and scheduling. An example is the Inspur ClusterEngine, which can manage and monitor the GPUs and schedule the GPU jobs, and is a suitable solution for heterogeneous HPC systems.

Chapter 3
Network Communication in a Supercomputing System

Mainstream network technologies for supercomputers include Ethernet, FC, and InfiniBand. FC is primarily designed for connecting storage devices. Due to its technical limitations, FC is only used widely in the storage domain. Both Ethernet and InfiniBand are open-network interconnect technologies. Ethernet focuses more on the versatility of the network protocols, with a unified protocol for both local area networks (LAN) and wide area networks (WAN). Therefore, Ethernet has been used widely in almost all network data transfer domains. With the data transfer speed of Ethernet gradually increasing to the same level or even above the FC, a lot of storage devices start to use Ethernet for interconnect as well. An example is the storage protocol ISCSI, which supports data transfer on Ethernet. On the other hand, InfiniBand is designed for making up the disadvantages of Ethernet and FC, as well as to meet the demands on performance and intelligence of the communication network and storage network. The performance of InfiniBand is far better than Ethernet and FC and includes intelligent features such as SDN. In recent years, InfiniBand has gradually become the dominating network technology in supercomputer systems.

We will mainly focus on the network features as well as optimization techniques of InfiniBand in this chapter.

3.1 Overview of the InfiniBand Technology

The standard of InfiniBand is designed by the InfiniBand Trade Association. It is a high-bandwidth low-latency network technology with an open standard. InfiniBand networks could achieve a high performance for a variety of network types, such as cluster communication network and storage network, with an extremely low cost on CPU loads. InfiniBand could be used both for building an internal high-speed

ASC Community, *The Student Supercomputer Challenge Guide*,
https://doi.org/10.1007/978-981-10-3731-3_3

network of a high-performance data center, and for building a high-speed data transmission network between data centers. InfiniBand architecture is introduced as follows.

3.1.1 Host Channel Adapter (HCA)

Host Channel Adapter (HCA) is the end node of the InfiniBand network, which could be installed in a server node or a storage node to connect the server or storage to the InfiniBand network. Between different HCAs, we could create a Queue Pair to achieve data exchange. Each queue pair consists of a Send Queue and a Receive Queue, and can be dynamically created as needed. By creating an information channel through Queue Pairs, we could avoid the communication dependency on OS, and enable the direct access to the HCA network resources (Fig. 3.1).

In contrast, in traditional networks, the applications need to rely on the OS to process network resources. The extra steps in the traditional approaches would affect the network performance.

3.1.2 Target Channel Adapter (TCA)

The Target Channel Adapter (TCA) is an adapter architecture that is primarily designed for embedded systems. A TCA does not provide standard APIs for applications and is only used in a small domain.

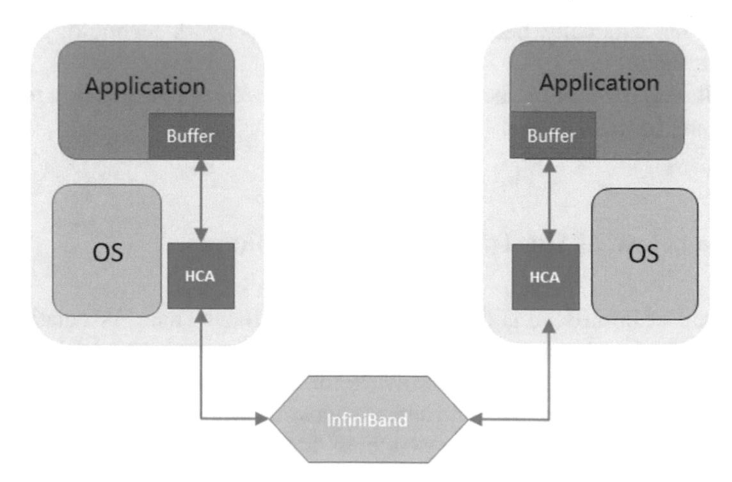

Fig. 3.1 HCA network accessing architecture

3.1.3 Switch

The InfiniBand switch is similar to traditional network switches. However, the design of the InfiniBand switch focuses more on the high performance and high performance cost ratio. Generally, the Cut-Through technique is used instead of the Store-Forward technique, so as to achieve a fast message pass as well as a low cost. InfiniBand switches also implement a linked-layer flow control mechanism to avoid loss of data packets. This feature is very important for InfiniBand to achieve reliable data transfer protocols.

3.1.4 Router

InfiniBand network routers could achieve direct interconnect of multiple InfiniBand subnets. In the past, we have not had a huge demand for InfiniBand routers. However, with the increase of the system scale, it becomes impossible to fit all the nodes into one subnet. Therefore, the InfiniBand router is becoming more and more important. Using an InfiniBand router, we could achieve data exchange among different InfiniBand clusters all around the world.

3.1.5 Cable and Connection Module

The InfiniBand standard supports both copper fiber and optic fiber. It also supports different link bandwidth (4×, 12×), and different speed (SDR, DDR, QDR).

By defining a complete system, InfiniBand could provide the most efficient network technology for data centers. Typical advantages of InfiniBand include:

- High performance. The newest FDR InfiniBand could provide a 56 Gb/s throughput in each port, and the Mellanox InfiniBand network card with two ports could provide a throughput of 100 Gb/s. A single port in an EDR InfiniBand switch can provide a bandwidth of 100 Gb/s (the first EDR InfiniBand switch was announced by Mellanox in June, 2014).
- Low latency. InfiniBand provides the lowest communication latency among different network technologies. The port-to-port latency is less than 0.7 ms for FDR InfiniBand. The latency advantage could significantly improve the performance of applications running on supercomputer platforms.
- High efficiency. InfiniBand has built-in support for a number of optimized network transport protocols, such as Remote Direct Memory Access, kernel bypass, Zero Copy, and Transport offload.
- Reliable and stable network interconnection. Using techniques such as full redundancy, lossless I/O, and multiple path failover, the high reliability of InfiniBand is of vital importance for dealing with critical missions.

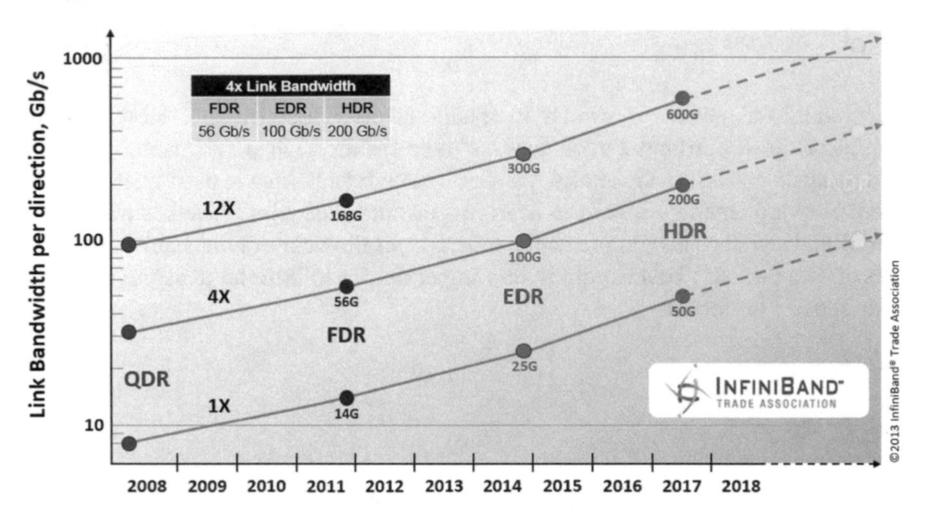

Fig. 3.2 InfiniBand roadmap

- Lower power consumption and cost with an integrated network solution. By integrating multiple networks (compute, management, and storage networks) into one unified network, we could cut down remarkably the network operation costs, achieving the best ROI for data centers.
- Data Integrity. InfiniBand provides Cyclical Redundancy Check in the communication of each port, providing the best accuracy of data transmission.
- Openness. InfiniBand has a complete ecosystem with open source software support.

In conclusion, InfiniBand provides the best available solution for applications that have a strict demand on network performance. Figure 3.2 shows the roadmap of InfiniBand. EDR products have just been released, and HDR (200 Gb/s) products will be employed around 2017.

3.2 The Current Status of InfiniBand in HPC

Clusters have already become the mainstream architecture of the supercomputers in the TOP500 ranking list.

As shown in Fig. 3.3, according to the TOP500 ranking list in June, 2014, more than 85.4% of supercomputers are clusters. The high scalability and high performance price ratios are undoubtedly the main driving force behind the replacement of MPP architecture with clusters. To improve the efficiency of the entire system, the inter-node communication is a key factor. The system requires a high-bandwidth, low-latency, highly scalable, and low-system-load network to improve the communication performance. As shown in Fig. 3.4, in June 2014

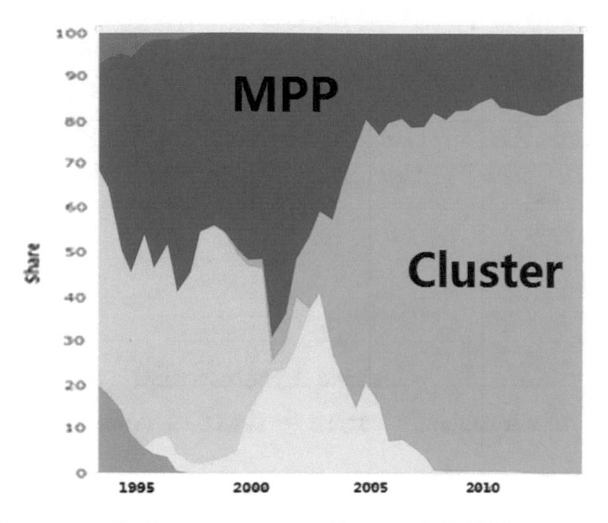

Fig. 3.3 Development trend of supercomputer architectures in TOP500

InfiniBand had become the most popular network employed in the TOP500 supercomputers replacing the position of Ethernet.

High performance and high performance/price ratio are undoubtedly the main reasons for the increasing ratio of InfiniBand in supercomputer systems. InfiniBand could bring a high system efficiency (system efficiency = real peak performance/ theoretical peak performance) of up to 99.8%. The real performance is the measured LINPACK performance of the system, while the theoretical peak performance equals to the sum of the theoretical performances of all the processors (assuming the

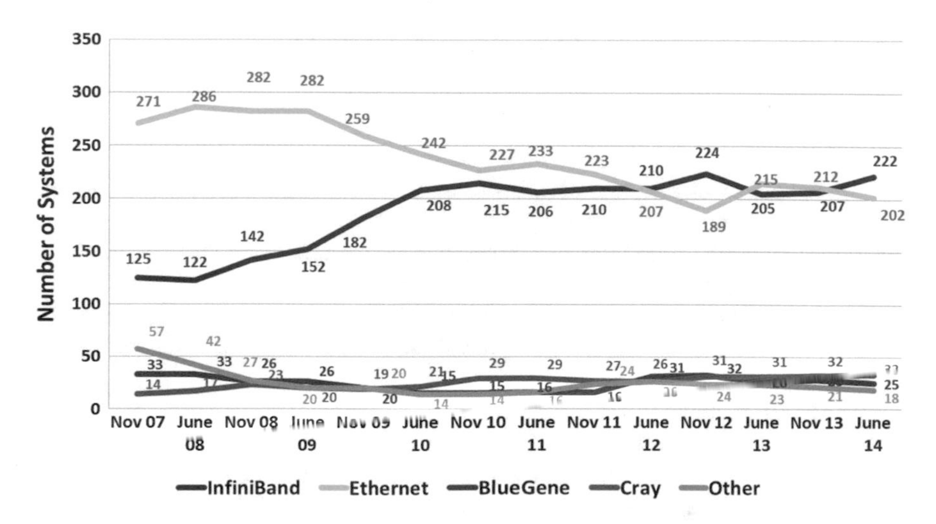

Fig. 3.4 TOP500 network interconnection trends

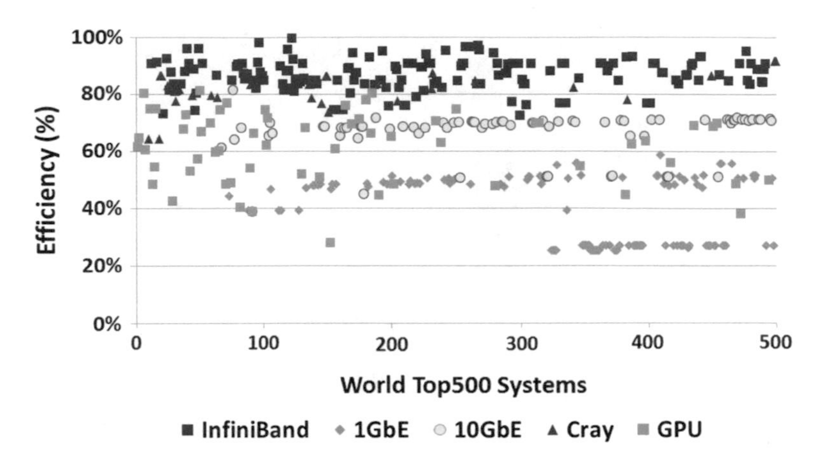

Fig. 3.5 System efficiency distribution

highest clock frequency and execution of the maximum number of floating-point operations in each cycle) in the system.

Figure 3.5 shows the system efficiency of TOP500 supercomputers applying different kinds of networks. The efficiency of systems employing InfiniBand is far better than systems applying Ethernet or private networks such as Cray's special network.

Though different people have different opinions about the efficiency of LINPACK, it is still an irreplaceable evaluation tool for supercomputers. If a supercomputer could not get a high efficiency of LINPACK, we would not expect a great efficiency for running other applications on that machine.

3.3 Core Technology of InfiniBand—RDMA

3.3.1 A Brief Introduction to RMDA

The throughput of message passing is not only determined by network bandwidth and latency. Remote Direct Memory Access (RDMA) is needed when we want to transform low latency into high accessing speed. Two types of RDMA are Offload and Onload. Different InfiniBand generally support different types. When we employ the Offload technique, it means that we have built the RDMA features into the InfiniBand chips. In contrast, the Onload technique means that we use CPU to process the network protocols.

The TCP/IP protocol is the most basic protocol of the internet, consisting of the Transmission Control Protocol (TCP) and the Internet Protocol (IP). This protocol is organized as four abstraction layers (network interface layer, network layer, transport layer, and application layer). As shown in Fig. 3.6, in a traditional network, to transfer the data, we need to copy the data through all these different

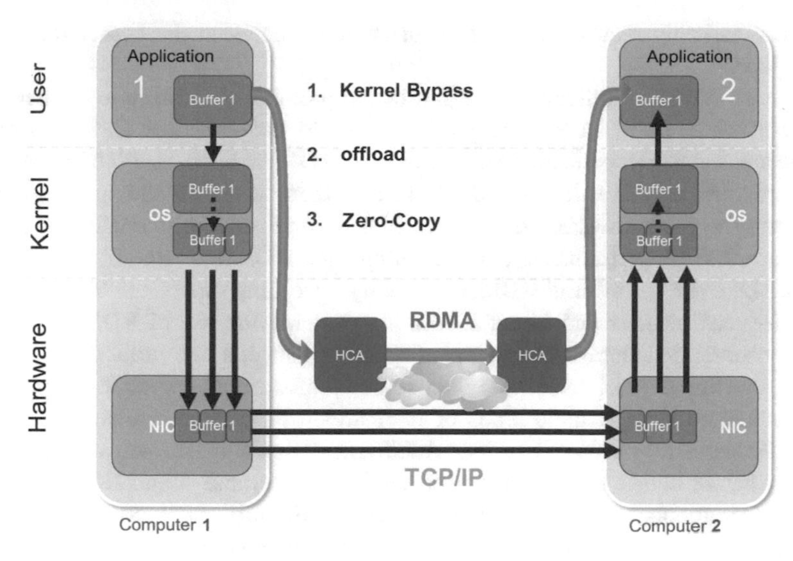

Fig. 3.6 Comparison between TCP/IP and RDMA

layers. The application needs to call OS services to achieve the corresponding network functions, leading to a low efficiency of data transfer. As a result, for supercomputer systems, we should not process communication using the CPU. The CPU's resources should only be used for performing computations.

RDMA enables data read and load between remote applications, by transmitting virtual memory addresses of remote applications in its message. The remote application only needs to register the corresponding memory buffer in its local network card. The advantage of the RDMA is that no CPU operations are needed for the remote node in that case. By using the type value (key value), an application can protect its memory when remote applications request random memory accesses.

The application that sends RDMA requests should set the right data type of the remote memory space it tries to access. Before the application starts the RDMA operations, the remote application needs to send the related info (memory address, size, and data type) to the application that sends RDMA requests.

3.3.2 Core Technologies of RDMA

Zero Copy HCA could exchange data with the application memory directly, avoiding the data copies between the application memory and the OS kernel.

Kernel Bypass When executing RDMA load/store operations, applications could directly pass its order to HCA without calling kernel functions. When adopting RDMA, the request could be passed from the user space into the local network card, and then from the local network card to the remote network card, which remarkably reduces the number of switches between the kernel space and user space.

Offload We could offload a large part of the protocol processing to the network card hardware, so that the communication would not consume CPU resources. There are some software implementations of RDMA, such as iWARP. As the software versions are still based on TCP, and there is no offloading of protocol processing to the hardware, the software RDMA performance is much worse than hardware RDMA performance (such as InfiniBand RDMA or RoCE).

RDMA over InfiniBand RDMA is rapidly becoming one of the basic features of high-speed clusters and server LANs. To promote the use of RDMA to interconnect server and storage nodes, OpenFabrics Alliance, a non-profit organization, works on the development of open source software OpenFabrics Enterprise Distribution to support three kinds of mainstream RDMA networks: InfiniBand, iWARP, and RDMA over Converged Ethernet (RoCE). The software includes drivers, kernel codes, middleware, user-layer interface, and supports a number of standard protocols, such as IPoIB (IP over InfiniBand), SDP, SRP, iSER, and DAPL. Meanwhile, it also supports multiple versions of MPI (message passage interface), different file systems (Lustre, GPFS, NFS or RDMA, etc.), and different OS (Linux, FreeBSD, and Windows).

Compared with other network technologies, the biggest difference of InfiniBand is that it could provide direct RDMA message passing for application layers, despite user applications or kernel applications. In contrast, in traditional networks, the access to the network has to go through the OS. In the environment of InfiniBand (as shown in Fig. 3.7), the interface between application layer and transport layer is Verbs, which is defined by the OFED Software Transport Interface standard. The application could send messages to the transport layer by calling the POST SEND Verb function.

Programs developed using Verbs can directly run on the InfiniBand network, and support RDMA naturally. Many high-performance applications are rewritten using Verbs, so as to achieve the best utilization of the networks that support

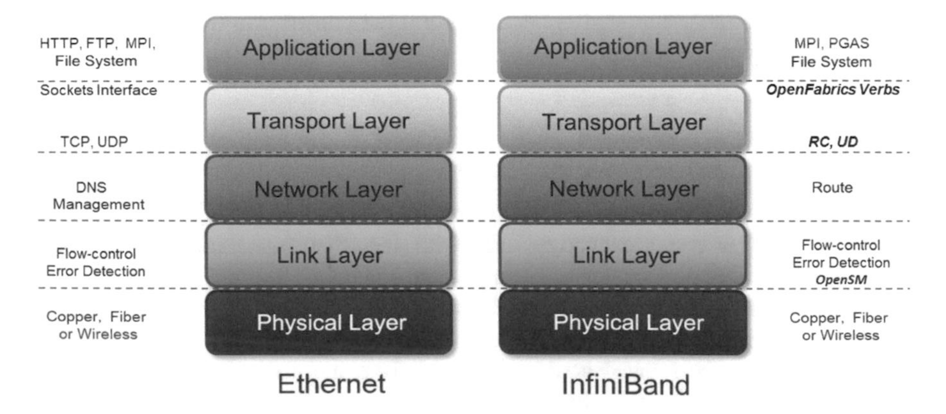

Fig. 3.7 Hierarchical architecture of the InfiniBand network

RDMA. However, to write Verbs programs, programmers need to know the basic rules of Verbs, the methods for creating Queue Pairs, establishing channels, and the methods for sending and receiving messages through the channel. To make it easier for programmers to use RDMA, OFED provides a series of ULPs (Upper Layer Protocols) to enable data transfer through RDMA instead of TCP. Each ULP provides two major interfaces: the upper interface for the applications and the lower interface for the transport layer. The upper interface provides standard interfaces that can be recognized by the applications:

SRP (SCSI RDMA Protocol) Using SRP, the SCSI file system could directly access remote storage using RDMA. No changes are needed for the file system itself.
SDP (Socket Direct Protocol) Socket applications could employ RDMA of InfiniBand to communicate without any changes.
iSER (iSCSI Extensions for RDMA) The iSCSI protocol enables remote access of block storage. iSER enables the iSCSI to run on the RDMA network.
IPoIB (IP over InfiniBand) This upper-layer protocol allows message passing inside the InfiniBand network or between the InfiniBand network and other networks using standard IP protocols. IPoIB supports TCP/IP, UDP, SCTP, and so forth. This protocol enables traditional applications to run on the InfiniBand network without any changes.
NFSoRDMA (Network File System over RDMA) NFS is one of the most widely used file systems. It provides file-level I/O. Traditional NFS could only run on TCP/IP. As an extension of NFS, NFSoRDMA could run on InfiniBand, and improve the file system performance using RDMA features. There are similar supports for other file systems, such as Lustre, GPFS, and so on.
RDS (Reliable Datagram Sockets) RDS was developed by Oracle, which provides the Berkeley Sockets interface, and supports the transfer of message from one socket to multiple destinations. By taking full utilization of the high-bandwidth and low-latency features of InfiniBand, the databases can use RDS to achieve the highest level of parallelism.
MPI This communication interface provides a full support of MPI functions for high-performance computing applications.

By using complete ULPs, the application can use one unified InfiniBand network to process different kinds of services, such as storage, network, and IPC services. The data center would not need multiple networks to achieve different services.

3.4 Optimization of HPC Applications Based on the InfiniBand Network

The MPI specification is standardized by the MPI Forum, and it has become the de facto industry standard for parallel programming. The newest version is MPI 3.0. Implementations of the MPI standard include MPICH and OpenMPI. MPICH is developed by Argonne National Laboratory and Mississippi State University, with a great portability. A number of other implementations, such as MVAPICH2, IntelMPI,and Platform MPI, are all developed based on MPICH. OpenMPI is an open source implementation of MPI maintained by a number of universities, research institutes, and companies.

In the HPC domain, parallel applications are typically developed based on MPI. Therefore, to optimize HPC applications, understanding the characteristics of MPI is critical.

3.4.1 MPI Communication Protocol

MPI communication protocol could be divided into two types: the Eager mode and the Rendezvous mode:

Eager mode

In this mode, the send process will actively send messages to the receive process, without consideration whether the receive process has enough buffer to receive the message. Therefore, this mode requires the receive process to pre-allocate enough buffer space for receiving incoming messages. The Eager mode has a relatively small initialization cost, which is quite suitable for sending of short messages that require low latency. In the Eager mode, we could send messages by using InfiniBand Send/Recv or RDMA to minimize the latency.

Rendezvous mode

Opposite to the Eager mode, in the Rendezvous mode, the send process would coordinate with the receive process for enough buffer space, which is suitable for sending long messages.

The Rendezvous and Eager modes are not limited to RDMA operations. They could also use Socket, RDMA Write, and RDMA Read. However, in Socket operations, the communication involves a lot of memory copy operations, which reduces the efficiency. Meanwhile, the communication operations cannot be overlapped with the computation operations. In contrast, RDMA Write and Read could achieve higher performance by using zero copy and kernel bypass, and can easily achieve communication and computation overlap. The process is shown in Fig. 3.8. The send process sends the control instruction "Rndz_start" to the receive process, then the receive process sends back the "Rndz_Reply" instruction to the send process, which contains the information about the receiving buffer and the key value. After receiving the "Rndz_Reply" instruction, the send process will directly perform the RDMA WRITE operation to transfer the data to the receive process. After sending the data over, the send process sends the "fin" instruction to the receive process to indicate that the data has been put into the buffer. The advantage of the Rendezvous mode is that we do not need to reserve memory space at the beginning. Therefore, it consumes less memory than the Eager mode.

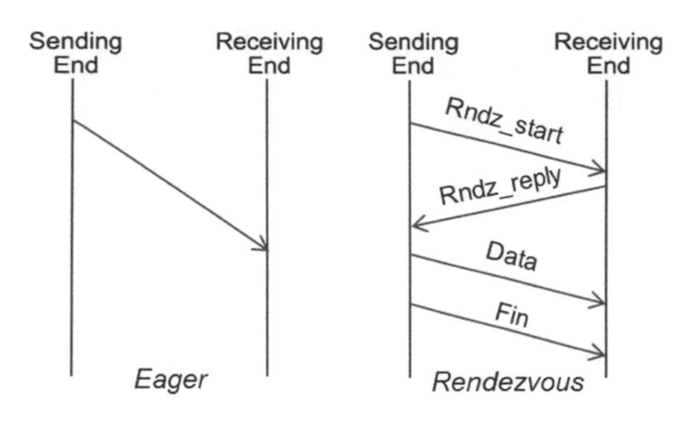

Fig. 3.8 The Eager mode and the Rendezvous mode

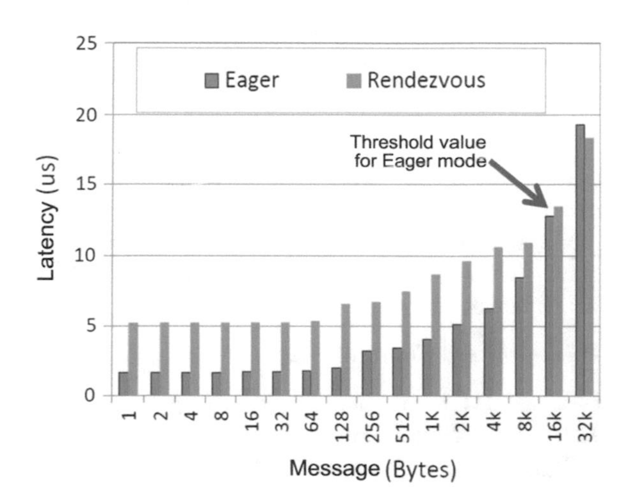

Fig. 3.9 Performance comparison between Eager and Rendezvous modes

The disadvantage is that the Rendezvous mode leads a longer latency. Therefore, the Rendezvous mode is more suitable for the transfer of long messages. As shown in Fig. 3.9, when the messages are smaller than 16 kB (the default threshold value for the Eager mode in MVAPICH2), the Eager mode provides a lower latency. However, when the messages are larger than 16 kB, the Rendezvous mode starts to show its advantages.

When running an MPI program, users could change the communication features to get a better performance by setting a suitable threshold value of the Eager mode, Fig. 3.10 shows an example. After changing the threshold value of the Eager mode, the latency curve changes accordingly.

3.4.2 MPI Function

As mentioned above, the underlying network protocol of MPI could affect the communication performance significantly. In addition to that, for the MPI functions

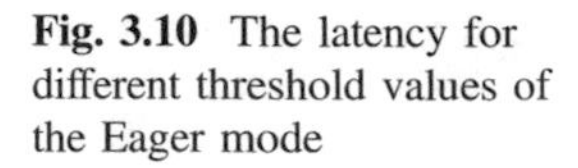

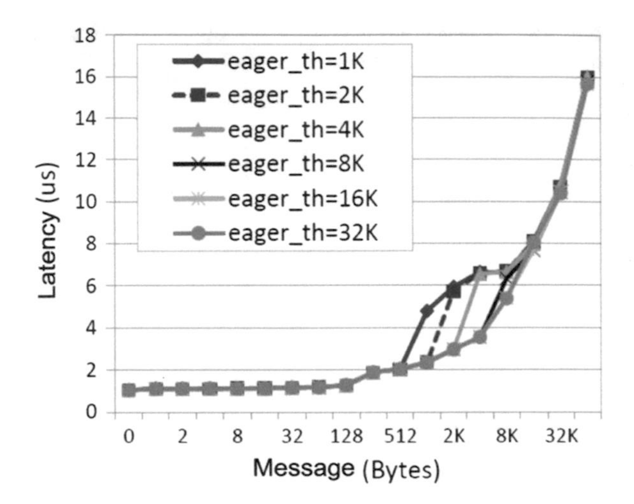

Fig. 3.10 The latency for different threshold values of the Eager mode

at the upper level, we can also perform another round of optimizations. There are two types of MPI functions, collective communication and point-to-point communication. Different MPI implementations have minor differences in these functions. Therefore, the optimization strategy should be different for different MPI implementations.

Point-to-Point Communication

MPI defines more than 35 point-to-point communication methods, including MPI_Send, MPI_Recv, MPI_Sendrecv, MPI_Isend, MPI_Irecv, MPI_Probe, MPI_Iprobe, MPI_Test, MPI_Testall, MPI_Wait, and MPI_Waitall.

Collective Communication

These include: MPI_Allgather, MPI_Allgatherv, MPI_Allreduce, MPI_Alltoall, MPI_Alltoallv, MPI_Barrier, MPI_Bcast, MPI_Gather, MPI_Gatherv, MPI_Reduce, MPI_Scatter, and MPI_Scatterv.

All the different MPI implementations have set different tuning parameters for these functions. To find the best parameter, the user should first run analysis tools to find out the latency of each MPI function to identify the most suitable parameters.

3.4.3 Performance Comparison of MPI on Different Networks

Two basic features of a network are the latency and the bandwidth. Latency rather than bandwidth is the most important factor when data size is less than 256 B, and vice versa. Figure 3.11 shows the measured bandwidth results of InfiniBand QDR

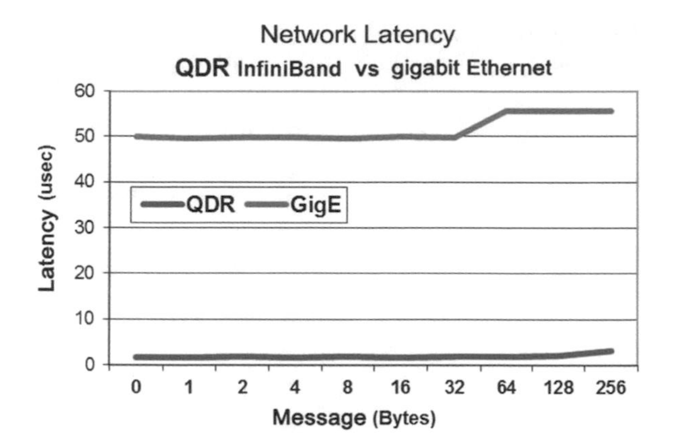

Fig. 3.11 Network bandwidth

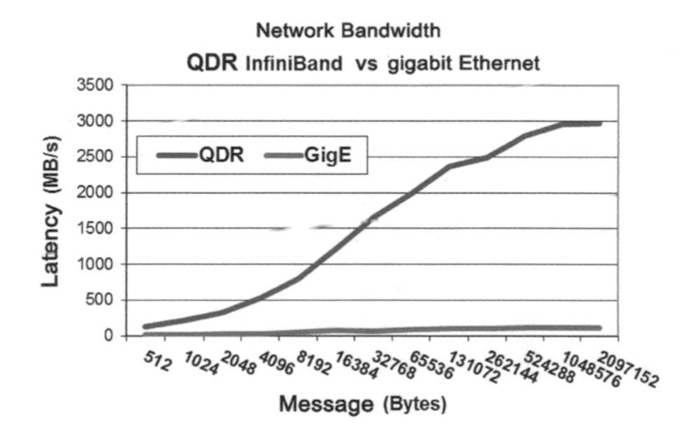

Fig. 3.12 Network latency

and gigabit Ethernet. With the increase of data size, network pressure becomes bigger and bigger, and so does the measured bandwidth of the two networks. According to the experimental results shown in Fig. 3.12, when the data size is 2 MB, the bandwidth of the InfiniBand QDR is around 3000 MB/s, which is much higher than the 100 MB/s bandwidth of the Ethernet. The latency of InfiniBand, which is around 1.2 ms, is smaller than Ethernet (50 ms) as well.

The above performance comparison is based on the QDR InfiniBand network. The more recent FDR InfiniBand would provide an even better performance. As shown in Fig. 3.13, the unidirectional bandwidth of the Mellanox Connect-IB network card can be as high as 12 GB/s, and the unidirectional bandwidth of the FDR InfiniBand network can reach 6.3 GB/s. The latency is smaller than 1 μs (Fig. 3.14).

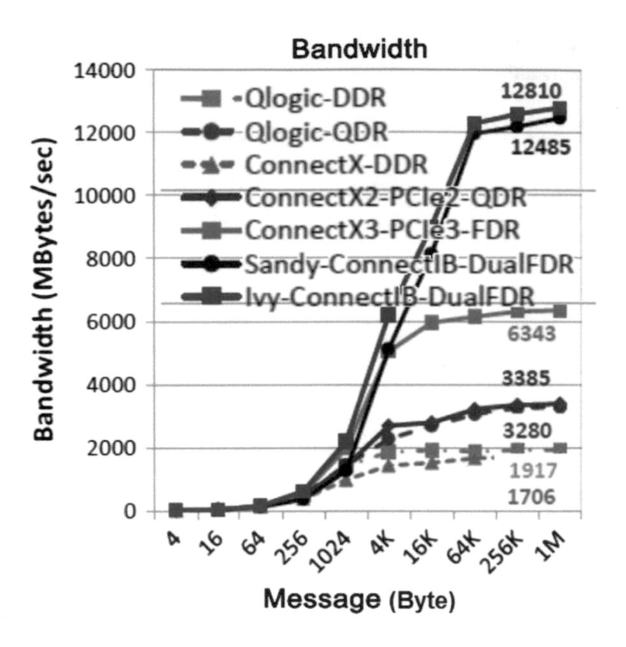

Fig. 3.13 Network bandwidth

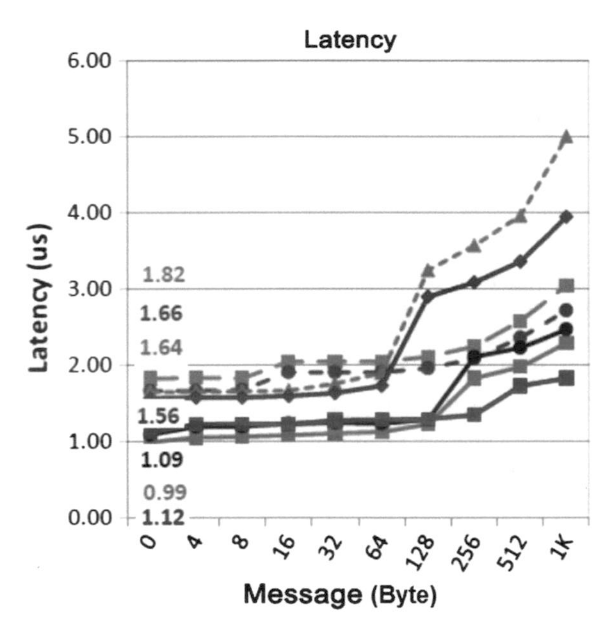

Fig. 3.14 Network latency

3.4.4 Application Performance

According to the performance testing report, InfiniBand has a higher bandwidth and lower latency. The ultimate metric for evaluating an HPC system is whether it can lead to a better performance of applications. Figure 3.15 shows the performance of

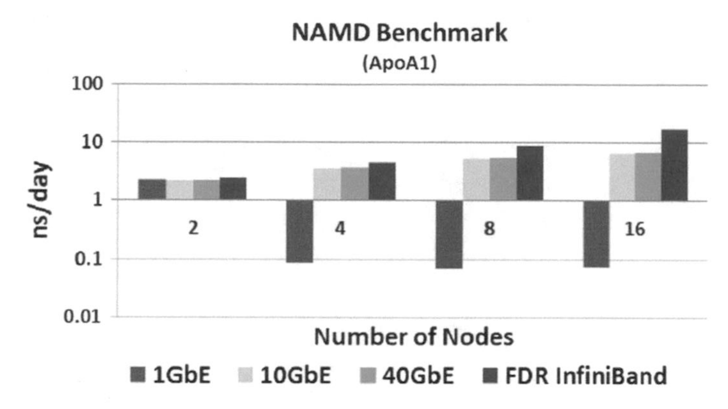

Fig. 3.15 Performance of NAMD on different networks

NAMD on two different networks. NAMD is one of the most popular molecular dynamics simulation software and Apoal is the standard performance evaluation test case for NAMD.

According to the experimental results, the InfiniBand network provides a better performance than the Ethernet network from the case of two nodes. When we increase the scale to eight nodes, the InfiniBand network leads two times better performance than the Ethernet network. With the increase in the number of nodes the performance benefits brought by using InfiniBand become more apparent. We can also see that from the scale of 4 nodes, the performance of NAMD on the Ethernet network increases very slowly. The 40-Gb Ethernet provides roughly the same performance as 10-Gb Ethernet, both of which are performing better than the gigabit Ethernet. However, when compared with the InfiniBand option, the Ethernet performance is much lower. Therefore, the InfiniBand network is a key technology for improving the performance and efficiency of HPC systems.

3.4.5 MPI Application Optimization

Different MPI implementations have different profiling tools. For example, OpenMPI has a tool called IPM, while Allinea can be used to analyze different kinds of MPI implementations. In our case, we use IPM to analyze the MPI applications.

Firstly we could use IPM to analyze the communication time of MPI. As shown in Fig. 3.16, when we run NAMD, the communication time of InfiniBand is much shorter than Ethernet, which could explain why we have higher overall performance on supercomputers employing InfiniBand.

We could further enhance the performance and scalability of programs by a careful MPI tuning. To identify our optimization target, we need to analyze the time taken by different MPI functions. An example is shown in Fig. 3.17. In this case,

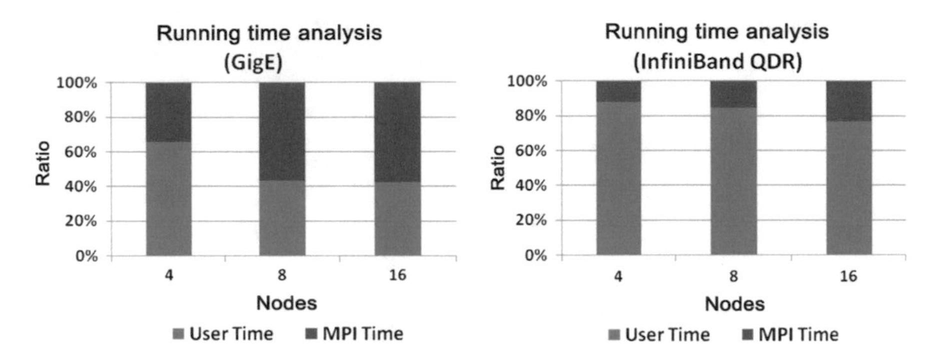

Fig. 3.16 NAMD running time analysis

running NAMD, MPI_Recv, and MPI_Iprobe are the most time-consuming functions. With the increase of the number of processes, MPI_Iprobe becomes the most time-consuming method. Therefore, we should first consider how to optimize these two functions.

To identify the most suitable MPI optimization parameters, besides analyzing the time taken by different functions, we also need to know the size of the messages passed by each function. In this case, as shown in Fig. 3.18, the IPM results show that most messages are in the region of 512 B–64 kB, and 256 kB–2 MB.

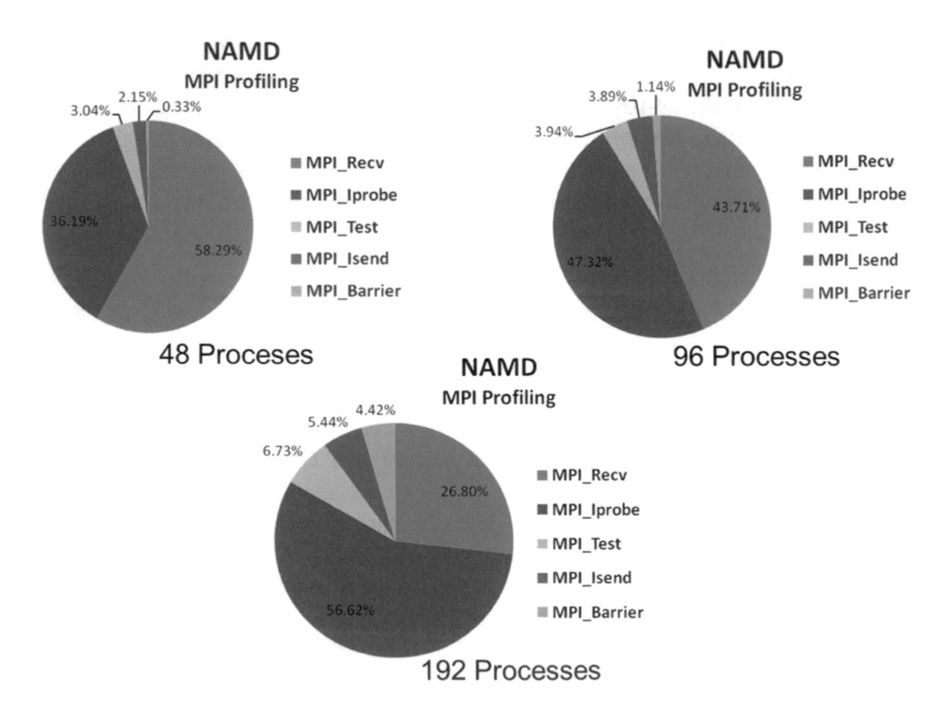

Fig. 3.17 NAMD MPI communication time analysis

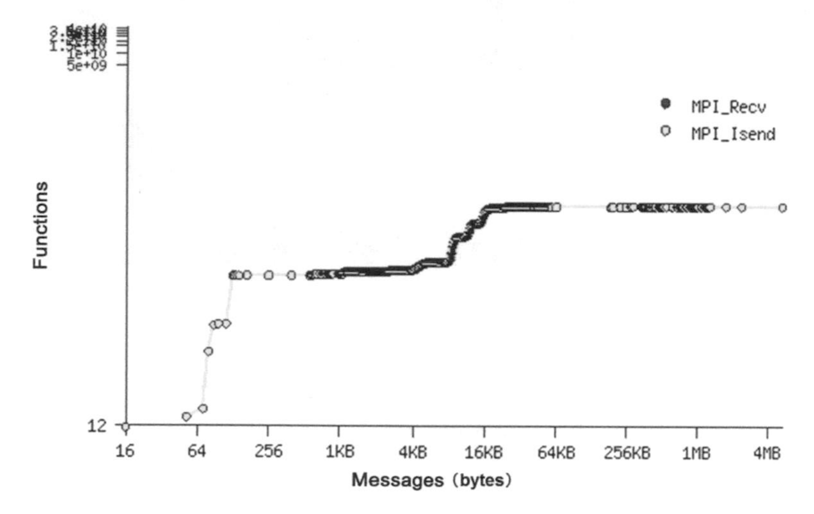

Fig. 3.18 NAMD MPI message size distribution

Therefore, based on the above information, we could consider optimizing MPI_Recv for receiving long messages, and we should pick MPI_Iprobe as the focus of our optimization.

As both MPI_Recv and MPI_Iprobe are examples of point-to-point communication, the most important parameters that affect the communication time would be Rendezvous and Eager parameters. After a careful tuning process, we find that when we increase the value of VIADEV_RENDEZVOUS_THRESHOLD to 50,000 B, the performance of NAMD on 16 nodes could be improved by 11% (as shown in Fig. 3.19). That is because when the message size is smaller than the threshold, the MPI functions would use the Eager mode to transfer the data, which reduces the communication latency and the run time of the program. However, we should also note that when we do a lot of Eager mode communication, we need to pre-allocate a lot of memory to ensure successful reception of the data. When we do not have much memory resource, increasing the portion of the Eager mode could also lead to memory issues and reduction of the running speed.

In conclusion, we should firstly use IPM to analyze the communication patterns of the application, and then perform a careful optimization to the corresponding functions and message sizes. For the applications that are not based on MPI, we should consider different optimization techniques. As shown in Fig. 3.20, the NAMD version that uses InfiniBand verbs could be 10% faster than the basic MPI version.

In the HPC domain, there are lots of new areas to explore for the network technologies. For example, when running MPI applications, we could call XRC or UD modes to perform point-to-point communication. Mellanox, a major vendor of InfiniBand network, has developed free open-source accelerating software MXM and FCA, to improve the speed of point-to-point and collective communications, and to further improve the scalability and performance by offloading more functions to the hardware.

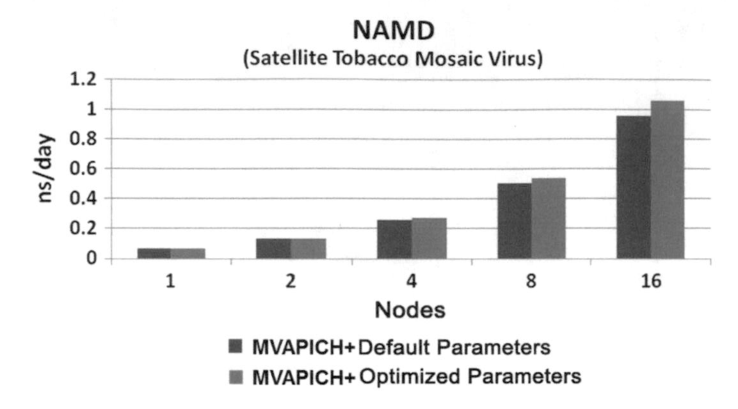

Fig. 3.19 Optimizations of the NAMD MPI function

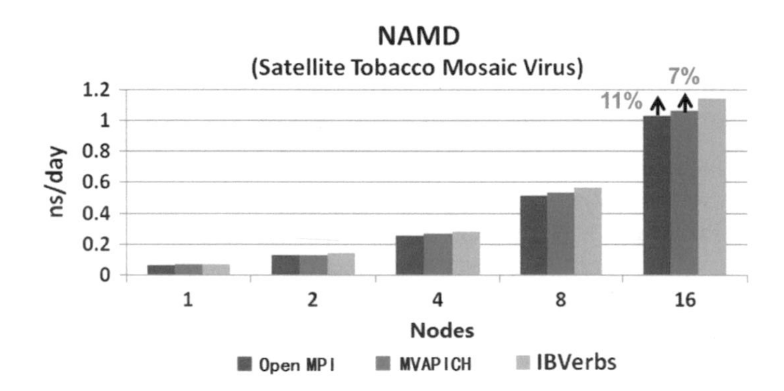

Fig. 3.20 Performance comparison of different NAMD versions

3.5 InfiniBand Accelerates Cloud Computing and Big Data

The RDMA network, such as InfiniBand, which is widely used in HPC, is now extending to more areas, such as big data, cloud computing, and enterprise applications. High-speed InfiniBand switch provides nano-second-scale network delay. Ethernet exchange chip can hardly achieve the line rate when transferring small packets. In contrast, the InfiniBand switch can achieve the line rate even when transferring small packets. The latency of InfiniBand network card is less than 0.7 μs, which is one-fifth to one-tenth of the speed of the Ethernet network cards.

In addition to the advantage on latency, InfiniBand provides inherent support for RDMA. This technology can lead to significant performance improvement for a series of applications in cloud computing (certain applications can be accelerated by several times). Examples that can all achieve significant performance improvement include:

- Big data Hadoop (inherent support for RDMA);
- Oracle, IBM DB2, Microsoft SQL server database;
- Block storage (support RDMA storage protocol iSER and SRP);
- Parallel file system (GPFS, Lustre, Ceph, etc.); and
- Enterprise applications, such as finance, futures, and animations.

Even if some applications do not support RDMA, the underlying storage or file systems with RDMA features can also bring performance benefits. Even if there is no RDMA support in the entire system, running traditional TCP applications on InfiniBand can still provide a much higher performance than the Ethernet.

Scalability InfiniBand is a software-defined flat network, and provides a good scalability. We can use InfiniBand to easily build a cluster with tens of thousands of nodes, which can be quite challenging for the Ethernet. Meanwhile, InfiniBand can provide the best scalability for the applications. In the big data era, the scalability of the applications is an important factor to consider when building a new system. InfiniBand provides congestion control, QoS control, protocol processing offloading to hardware, and hardware virtualized SR-IOV. These features can significantly improve the scalability of the applications running on top of InfiniBand.

Performance Price Ratio The actual performance price ratio of a system = (100% − the percentage of the network price) × overall efficiency of system.

Taking the real data as an example, assume that the gigabit Ethernet's cost takes about 10% of the entire system's purchase cost, and the overall efficiency of the system is 60% (based on the data in the TOP500 list). The performance price ratio is (1 − 0.1) × 0.6 = 0.54 (see Table 3.1).

If we adopt the most advanced InfiniBand, the purchasing cost of which takes 20% of the overall cost, and the overall system efficiency can achieve 90% (based on the data in the TOP500 list), then the system performance price ratio will be (1 − 0.2) × 0.9 = 0.72 (see Table 3.1). Therefore, although the InfiniBand is more expensive than the Ethernet, it provides a better performance price ratio.

Support Cloud Computing Platform The InfiniBand network supports all the mainstream popular virtualization software, such as VMWare, OpenStack (KVM, Xen), and Microsoft Hyper-V. On the mainstream virtualization platforms, we can achieve high-performance virtualization of the InfiniBand network card, and integrate multiple networks into one. Therefore, we no longer need the FC network in a cloud computing environment.

Table 3.1 Performance price ratio

The percentage of the network price (%)	Efficiency of system			
	60%	70%	80%	90%
10	0.54	0.63	0.72	0.81
20	0.48	0.56	0.64	0.72
30	0.42	0.49	0.56	0.63

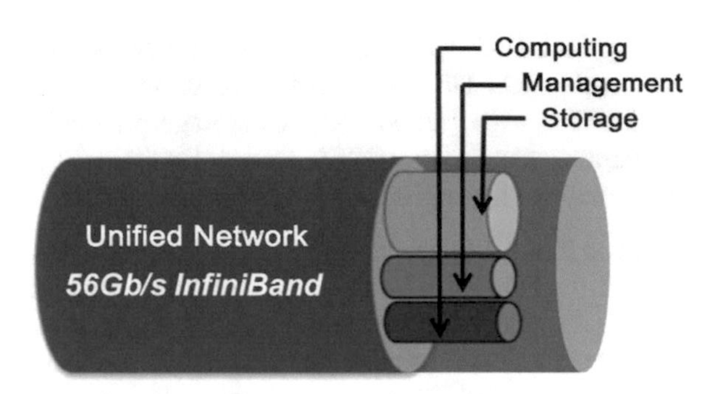

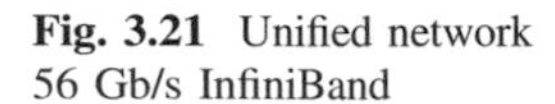

Fig. 3.21 Unified network 56 Gb/s InfiniBand

A unified InfiniBand network can satisfy the various communication needs for computing, management and storage (see Fig. 3.21). Therefore, we can use a simplified cloud network, and reduce the purchase and operation costs.

In recent years, there have been more and more cloud computing systems using the InfiniBand network, due to the low power consumption, high application performance, high scalability, high performance price ratio, and simplified network structure. Oracle's public cloud, which is based on the InfiniBand network, is now supporting 7.6 million global users, and 5000 PaaS customers. During the fourth quarter in 2013, there was an increase of 500 large-scale SaaS customers. There are 16.5 billion transactions happening on the platform, which connects to 90-PB storage through the InfiniBand network. The public cloud platform of Microsoft also uses InfiniBand to improve the performance and scalability of the system. The improvement on CPU, memory, storage, and application, all need the support of RDMA networks. More distributed application of big data requires the support of the high-speed network of RDMA. IBM also uses the InfiniBand network on the SoftLayer, leading to a faster network speed than Amazon and Google. The InfiniBand network helps the platform of IBM to support users with compute-intensive and communication-intensive applications, and to achieve a better performance price ratio. With the InfiniBand network integrated into cloud computing platforms, new kinds of applications will emerge. These new applications will be able to make full use of the capabilities of the cloud, and promote more demands and utilization of the cloud services.

Therefore, the InfiniBand network is not just limited to HPC. With its simple network structure, high network performance, low power consumption, high scalability, and more complete ecosystem support, the InfiniBand network is becoming the best solution for the current cloud computing centers.

Chapter 4
Building the Environment for HPC Applications

Six decades have passed by since the emergence of the first digital computer, and the computer industry has experienced several major technology revolutions. From the initial batch computer to the current parallel processing system based on SMP, MPP, NUMA, and clusters, computer systems have changed greatly. With the popularity of multi-core CPUs, parallel applications have gradually become the mainstream. Therefore knowing how to build an environment for parallel applications has become an urgent issue.

The current mainstream platforms for running HPC parallel applications can be divided into two categories, the cluster system based on shared storage systems and the cluster system based on distributed storage. Note that the storage refers to the memory rather than external storage. The distributed storage cluster system is composed of multiple homogeneous or heterogeneous computer nodes connected by a network to cooperatively accomplish a specific task. Because of its high performance-cost ratio, the distributed storage cluster system is applied in most cases.

The software for building an HPC application environment mainly consists of the operating system and the parallel environment. The operating system can be divided into two major families, Unix and Windows; Unix is the dominating solution that accounts for more than 90% of all the systems. Within the Unix family, the Linux operating system is in the majority. The Linux system has several advantages:

- A low cost. Most Linux distributions are free to use, with a charged service in some cases, which significantly reduces the software costs.
- Free and open. Being free and open are two most significant features of Linux. Linux developers have evolved from only a handful of people to tens of millions of users worldwide at an amazing speed. One major reason behind this miracle is the broad space provided by the open-source mechanism. Users around the world provide a huge number of tests for the system, enabling a timely fix of system bugs. Meanwhile, global developers are bringing their contributions to

ASC Community, *The Student Supercomputer Challenge Guide*,
https://doi.org/10.1007/978-981-10-3731-3_4

make a better Linux system, making it more and more efficient, and more and more convenient to use. For HPC development engineers, open source means that they can fully understand how the program is operating in its own operating system. Based on this observation, they can make targeted tuning to improve program performance. They can even modify the operating system to make it more suitable for hardware and applications.

- Efficient. Due to the separation of the Linux kernel and the graphical interface and the initial purpose of designing the system its stability is much better than other systems. Linux is a single-kernel structure, which makes the efficiency of the Linux kernel higher than microkernel operating systems. Thus, the Linux system is ideal for running HPC applications. The user's choice has also proven this point. In the Top500 list, among the world's fastest 500 clusters, more than 90% are using the Linux system. This has fully proven the strength of Linux as the HPC operating system.

For the parallel environment, MPI is nowadays the most widely used solution, due to its good portability, powerful functions, and a high efficiency, and so forth. There are many different kinds of free and efficient MPI implementations. Almost all of the parallel computer manufacturers and operating systems provide the support for MPI, which is unbeatable for other parallel environments. Despite its late appearance in 1994, MPI has absorbed the advantages of other various parallel environments and has taken many different factors, such as the performance, functionality, and portability, into account. It has caught up with the golden era of parallel computing development, spread fast within a few years, and soon became the actual standard of the message passing parallel environment.

With the increase of supercomputing applications, and the higher demands for computing power, the CPU architecture gradually became an uncompetitive solution. To solve this problem, people proposed the idea of heterogeneous computing. The term "heterogeneous" refers to the computation devices that are different from traditional CPUs, such as GPU and MIC. These new devices generally provide significantly higher parallel performance than traditional CPUs. GPUs were originally used for graphic processing, but not for high performance computing. However, the GPU architecture is quite suitable for the highly parallel computations. In the old days, to utilize GPU devices, people need to first transform the original general problem into a graphic or visualization problem. Therefore, the programming part involves a lot of difficulty, and GPU-based high performance computing is not very popular.

In 2007, NVIDIA released CUDA for solving the GPU programming issues. At the hardware level, the GPU architecture has been optimized for high-performance general-purpose computing. At the software level, this architecture developed CUDA language, which is similar to C language and provides developers with an easy-to-use method of using the GPU for parallel computing.

In late 2012, Intel formally released Inter Xeon Phi. MIC adopting x86 instruction set; the program could be transplanted to MIC without any change.

Prior to this, the heterogeneous development environment also included ARM, FPGA, and other non-x86 architecture environment. However, they were too complex and could not be combined with the existing x86 architecture environment due to a huge difference. For the case of GPU, CUDA has provided a good interface to program the GPU, that combine the computing capabilities of both CPUs and GPUs, thus making the heterogeneous computing a practical solution.

The heterogeneous system does not abandon the usage of CPU. Instead, it combines CPU with the new computing hardware for parallel computing. Based on the co-processor mode, either GPU or MIC adopts PCI-E interface to connect with existing x86 nodes. Therefore, we can add part of the computing power without abandoning the existing cluster. As a result, we see more and more heterogeneous systems in the Top500 list. In the Top500 list of November 2012, there are 62 clusters adopting the heterogeneous co-processor mode.

On the hardware side, compared with the traditional cluster, the heterogeneous cluster has no revolutionary changes. We usually add heterogeneous computing cards into the existing nodes in the homogeneous cluster to build a heterogeneous cluster. As for the software, however, there may be a greater change. In terms of the different heterogeneous hardware approaches, we need to use different software environments and write the corresponding codes for the CPU and GPU/MIC. More importantly, due to the different structures of computing devices in the heterogeneous cluster, we have to use different ways to solve different problems in different situations. We cannot use the traditional thought of CPU programming to code on GPU/MIC devices, and new programming thinking should be focused on the high parallelism. This shift in the way of thinking is sometimes more different than in the hardware mode. However, only when traditional serial thinking changes into the heterogeneous parallel thinking can we fully take advantage of heterogeneous hardware, and CPU + GPU or CPU + MIC collaborative computing, to reach the best performance.

4.1 Applied Environment for CPU Parallel System

4.1.1 Hardware Environment

The most commonly used parallel environment is the distributed-storage enclosed cluster system. The cluster consists of multiple compute nodes, each of which is a shared memory computing system. Generally, each node in the cluster adopts the same configuration in order to facilitate the management and the performance optimization. The node is usually a standard rack-mounted server with two-way or four way multi-core server-version CPUs (such as Intel Xeon series) and each core has 4 GB or more memory. Nodes are interconnected by Ethernet cable (gigabit network cable or 10-Gb network cable) or InfiniBand network cable through the corresponding switches. All nodes belong to the same intranet, and each node has unrepeated IP address of the intranet. Clusters are usually physically isolated from external network to ensure the security.

For detailed information regarding the methods and details for building specific hardware environment, please refer to the corresponding chapters in the previous chapters.

4.1.2 Software Environment

4.1.2.1 Operating System

In general, Linux is the operating system that is widely applied in parallel computing applications. Among all the Linux distributions, RedHat is known for its stability, and has been widely used in commercial clusters. However, RedHat requires the license fees for maintenance and the after-sales service. As a result, some of the most cost-sensitive customers give up the RedHat Linux distributions. One variation of RedHat is CentOS. Using CentOS, customers can get a similar experience to RedHat without any charge. Since CentOS is not officially supported, the patch upgrade may not be as good as the official RedHat, and the after-sales service is also not provided. In addition to RedHat, another major commercial Linux distribution is the SLES (SUSE Linux Enterprise Server) from Novell Incorporation, which also requires the users to pay for the usage. Note that the parallel application clusters all use the server-version operating systems. These systems have been optimized for server applications, making its performance and stability more suitable for HPC applications. Since some desktop-version operating systems are for individual users, they conduct some optimization on the desktop and display but lack the optimization on network connection and system services provision. In addition, in security fields, people would often use the FreeBSD operating system based on Unix, which was designed with a focus on improving security features.

4.1.2.2 Parallel Running Environment

Since the hardware environment mentioned above adopts the distributed memory architecture, the parallel software environment requires network communication using message passing. MPI (Message Passing Interface) is the most widely used parallel programming environment with the advantages of easy programming and strong portability. A most important implementation of MPI standard is MPICH, which was jointly developed by Argonne National Laboratory and Mississippi State University. It is a specific implementation of MPI, and can be freely downloaded from the Internet.

The installation steps of MPICH parallel programming environment are:

(a) Decompression and installation. First, download the installation package from the Internet, and use the appropriate command to extract the package according to the suffix. Enter the MPI decompression directory, and use ./configure and make the command to configure and automatically compile.
(b) Decentralization. In order to run MPI programs on several different machines, we firstly need to decentralize to the machines that have started MPI, that is, we authorize the machine that has started MPI to access other machines. For simplicity, we can create the same account name on each node in the entire cluster so that MPI program can run under the same account. After completing decentralization setup, we can use the same account name on any node without password. Thus we can start the MPI program at any node, and automatically run the program without entering a password.
(c) Operation. Programs at different nodes can be placed in different directories. However, in order to facilitate management, it is recommended that the program is placed in the same directory based on the shared storage. Which nodes to use and how many MPI processes should be created are determined by the configuration file. Every application can use different node configuration files, so as to choose different solutions according to the applications, and to avoid low efficiency when the computing resource is not fully used.

Within the node, since it is based on the shared memory model, one can choose to use MPI, OpenMP or pThread to achieve parallelization within the node. pThread is a multi-thread library, and is the basic multi-thread library under Linux. We can see some ported pThread libraries in other operating systems as well. OpenMP is based on pThread. Its underlying functions are implemented using the pThread library. By the way of inserting compiler directives into the source code, OpenMP can shield the underlying details of the parallel program development, and becomes the most convenient way to enable the parallelization. However, unlike pThread, OpenMP library does not provide interfaces for controlling some of the underlying details. Thus in some cases, its performance may be slightly lower than pThread.

4.1.3 The Development Environment

Traditionally, the HPC parallel environment is built based on the application itself, which means that the underlying architecture should be in accord with the application. Ideally, the development and the execution of the application should be based on the same platform. However, in practical cases, while the application is generally developed on one platform, it could be running on a number of different platforms. Therefore, it is generally a good strategy to choose the development environment with the same architecture and software and hardware environment as the running environment. In most cases, the running environment would have a larger scale than the development environment. As different scales could lead to different performance behaviors, we should emulate the target environment to perform the testing, so as to avoid apparent performance degrade during the deployment stage.

The development environment not only contains the same software environment as the application running environment, but also includes some software development tools (such as software compilers, debuggers) and optimization tools (such as high-performance math libraries, tuning tools).

4.1.3.1 Compiler and Debugger

In terms of the compiler, users typically use the Intel Compiler XE Collection, including C/C++, Fortran, and other language compilers. Generally speaking, Intel compilers will produce a good performance for the application, especially when the application is running on the platform of Intel CPUs. However, as Intel compilers are not free, the users who care about cost more than performance would prefer free alternatives. GNU/gcc is a free compiler collection from GNU, including C/C++, Fortran, and other language compilers. The GNU compiler has corresponding implementations on various operating systems and platforms, and has been widely used, especially in the Linux system. In the Linux system, the GNU compilers are installed by default.

The Linux system includes the running environment support for pThread libraries. All popular compilers support OpenMP syntax and pThread function declaration, and come with OpenMP and pThread runtime libraries. Therefore, after the installation of the compiler, the system can support OpenMP and pThread libraries.

MPI runtime libraries and the compiler environment are included in the MPICH installation package. In the MPICH compilation mode we can choose to use other compilers such as an Intel compiler. If no specification is given the system will use the default GNU compiler.

The debugger generally adopts a free GNU/GDB, Intel IDB, or TotalView debugger.

GDB is usually a default component of the Linux system, and is tightly integrated with the system. GDB can be applied to a wide range of applications, and can support a lot of different systems. Since it is based on the console interface, GDB is suitable for remote debugging and debugging scenarios without a graphical interface environment. GDB supports common functions, such as breakpoints setting, code viewing, stack checking, and thread adhesion. GDB is powerful and is widely used to debug in various environments.

IDB is the commercial debugger from Intel Incorporation. It has debugging tools with a command line interface as well as a graphical interface supported by JAVA. Therefore, IDB is relatively simple to learn and use. Since IDB is fully compatible with GDB commands, GDB users can switch to IDB without too much effort. In addition to the graphical interface, IDB has also improved some weakness of GDB. Therefore, IDB becomes an advanced choice.

TotalView is a full-featured and resource-level graphical debugger that provides complex control on the application for the developer. It can provide advanced supports for the large and complex programming development using information inspection, multiple threads, OpenMP, and other parallel programming techniques. These parallel programming techniques are able to be used on distributed-memory and shared-memory multi-processor computers. TotalView debugger has an extensive and complete C/C++ debugging feature set. The advantages of the TotalView debugger lie in its powerful graphical interface and the support for multi-process, especially for the parallel cluster system. Therefore, TotalView is generally considered the most powerful tool for debugging parallel applications.

4.1.3.2 High-Performance Math Function Library

The existing mathematical function library can simplify the process for the application development so as to avoid repeated work and to make the program concise. Since the existing math library is formed after a large amount of development, optimization, and testing, in most cases the use of existing math library will bring higher performance and better reliability than writing one's own code. Therefore, if some classic algorithms are used in the program, it is better to adopt the existing high-performance math library to improve the engineering quality and the performance of the codes.

Intel MKL (Math Kernel Library) offers many math routines that are highly optimized and parallelized, and is used for the performance-demanding applications, such as scientific, engineering, and financial fields. MKL is a toolkit component of Intel C++ and Fortran compilers (professional edition), it is also available as a separate product. For the current multi-core x86 platforms, MKL has made deep and comprehensive optimizations such as making full use of the CPU vectorizing instructions. MKL supports C and Fortran interfaces for multi-core CPU. It contains thread-safe features and has an excellent scalability on multiple cores. It can automatically conduct the processor detection, and select and call the most appropriate binary code for the processor. MKL includes a linear algebra math library (BLAS and LAPACK, ScaLAPACK, Sparse Solvers, Fast Fourier Transform) and a vector math library (such as power, trigonometric, exponential, hyperbolic and logarithmic functions, random number generator, etc.). Similarly, we MKL is a paid for service, but the distribution of its runtime library is free.

FFTW (Faster Fourier Transform in the West) is a standard C language program set for the fast calculation of Fourier transforms, which was developed by Frigo and Johnson from MIT. It can be used to calculate one-dimensional or multi-dimensional data as well as any scale DFT (Discrete Fourier Transform). FFTW also contains parallel versions for shared and distributed memory systems, and can automatically adapt to the compute nodes and fully utilize the cache, memory size, and the number of registers. FFTW is usually faster than other open-source programs for Fourier transform. The FFTW code generator adopts object-oriented design techniques and is written with the object-oriented language Caml. It can automatically adapt to the system hardware with high portability. For a model of any size, FFTW generates a plan through which it executes various operations and completes the conversion between various modes. Its internal structure and complexity is transparent to the user and the operation speed is fast (it is suitable for the internal compiler and code generator of various machines. Via AST, the code will be generated and self-optimized during the operation without occupying compile time. It uses hierarchical storage technology). FFTW is generally favored by a growing number of scientific and engineering computing developers, and provides practical large-scale FFT calculations for many application domains, such as quantum physics, spectroscopy analysis, audio and video streaming signal processing, oil exploration, earthquake prediction, weather forecasting, probability theory, coding theory, and fault diagnosis of medicine.

BLAS (Basic Linear Algebra Subprograms) is an API (application programming interface) standard for publishing the numerical library of basic linear algebra operations (such as vector or matrix multiplication). The subprogram was originally released in 1979, and is used to create a larger numerical package (such as LAPACK). In the field of high performance computing, BLAS is widely used. For example, the computing performance of LINPACK is largely dependent on the performance of the BLAS subroutine DGEMM. To improve the performance, the hardware and software vendors have highly optimized the BLAS interface according to their product.

LAPACK (Linear Algebra PACKage) is a function set that is written in Fortran and is used for numerical computations. LAPACK provides a wealth of tools that can be used for: the solution of multiple linear equations; least-squares solution of equation set of linear systems; calculating the eigenvector; singular value decomposition; and the Householder transform for matrix QR decomposition. NetLib also provides LAPACK95, a simplified API with Fortran 95. LAPACK is published with the authority of BSD.

The above mathematic libraries are all free to use, can achieve better performance than other similar free libraries, and have been used widely by the scientific computing community.

4.1.3.3 High-Performance Tuning Tool

A high-performance tuning tool is used to help identify the bottlenecks in existing programs through sampling and analyzing the code, and to assist developers in improving performance. Since the programs of HPC applications are usually very complex and time-consuming, performance optimization staff always encounter issues including how to quickly and accurately find the program hotspots and how to improve the performance accordingly. Speculating the hotspots according to the understanding of the source code is largely dependent on the person's experience and is very inaccurate. Although manual sampling to test procedures can objectively reflect the true performance of the situation, this method is time-consuming, requires modifications to the code, and could miss the real hotspots in many cases. Using a high-performance tuning tool can avoid such problems. Through the instrumentation of the source code or the binary file, we can accurately, objectively, and comprehensively test the program and find the real hotspots without modifying the source code. After analyzing these testing results, the optimization staff can accordingly optimize the performance.

GNU gprof performance debugging tools are included in the Linux development and debugging suite. By adding "-pg" compiler options when compiling, gprof will automatically add an instrumentation when the function is called. When the program is finished, it can output the running time and call relationships of each function, which makes the performance optimization process much more convenient. Printing out the consumed time of each function during the running process can help programmers to identify the most time-consuming functions. Generating function-call relationships including the call times can help programmers to analyze the program workflow.

Intel VTune Performance Analyzer is commercial software. Simply by rerunning the program with VTune, even without recompiling, we can automatically obtain the operating information of programs. In addition to the function consuming sampling, VTune can also do code-level sampling, even assembler-statement-level sampling. Furthermore, VTune can sample different hardware events for the users to conduct analyzing and testing. In addition, VTune can better support multi-thread and multi-process analysis and testing work. Therefore, it is particularly suitable for tuning the high-performance computing parallel applications.

4.2 The Application Environment of CPU + MIC Heterogeneous Parallel Systems

4.2.1 The Hardware Environment

In the heterogeneous environment, GPU and MIC adhere to the existing system in the form of an additional card, so that we do not need to change the overall hardware structure. We only need to add intra-node co-processors or replace the current node by the node with co-processors.

Since there is no difference between GPU and MIC in the system hardware structure, we will take MIC as the example to introduce the hardware environment of heterogeneous systems. Heterogeneous cluster architecture is shown in Fig. 4.1

In addition to the issues that we need to consider when building traditional high-performance hardware environment, three special issues should be noted when adopting heterogeneous co-processor architectures:

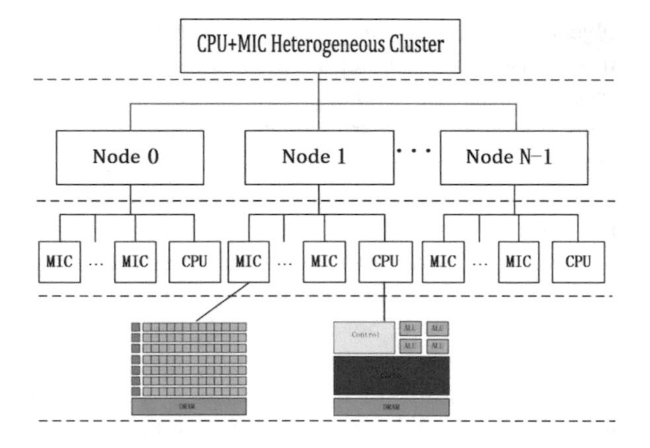

Fig. 4.1 The structure of the heterogeneous hardware environment

(1) We may not be able to add GPU or MIC cards into the traditional compute node, because there may be no available or appropriate PCI-E slots. For current cases, the nodes need to have one or more spare PCI-E (2.0 or 3.0 version) x16 slot(s) to support heterogeneous accelerators.
(2) Adding co-processors greatly increases power consumption requirements. Generally one extra co-processor requires a 300 W power supply, which is not a small overhead. Even so, through the analysis of performance per watt, we find that the accelerators provide far better power efficiency than that of traditional CPUs. Therefore, to achieve same level of computing power, using the co-processor can save more energy.
(3) Adding the co-processor also results in a higher requirement of heat dissipation of nodes. When building a heterogeneous cluster, we have to consider whether the node can support the co-processor or not, and also need to pay attention to the heat capacity of the node.

Please refer to former related subsections for more details.

4.2.2 The Software Environment

Since the hardware environment is heterogeneous, the building of a heterogeneous system software environment includes two parts, the general environment and the specific software environment that supports the added new hardware components. The general part corresponds to the traditional program and is the basis of the specific environment for heterogeneous accelerators. In the specific environment for accelerators, the drivers for new hardware components are also built on the traditional environment.

4.2.2.1 General Software Environment

The general parts of the software environment are similar for both homogeneous clusters and heterogeneous clusters.

For the parallel environment, we generally need to include the general components, such as MPI, and OpenMP. InfiniBand is the common network interconnection method and is supported by the existing heterogeneous environments. Therefore, the general software environment can also add the drivers that support InfiniBand according to the requirements of the hardware.

4.2.2.2 The Specific Software Environment for MIC

Similar to GPU, the software environment for MIC is also divided into the driver and the run-time library. The driver's role is to generate links between software and hardware to ensure that the operating system can successfully utilize the hardware device. The driver is closely related with specific hardware. Therefore, we should download the exact version that corresponds to the hardware configuration. Note that the latest driver does not necessarily provide the best performance. The latest drivers are generally optimized for the latest devices, and the performance of previous generation devices might degrade. In general, we should use a relatively new and official version of the driver. Although some beta-version drivers might bring performance benefits, we generally do not use them for industrial applications due to stability considerations.

A runtime library contains a collection of libraries that need to be used when running the MIC programs. These libraries call the interfaces provided by the driver to perform MIC-based computations.

Similar to GPU, the driver must be installed on each node, while the runtime library can be installed only on the shared directory.

For the parallel running environment, MIC supports the standard OpenMP and MPI libraries. However, if we want to take full advantage of the characteristics of the MIC (e.g. regard MIC card as a separate node), we need to install and use Intel MPI.

4.2.3 The Development Environment

The development environment should be built according to the running environment, with the same software/hardware configurations and scale. However, because of cost limitations, the scale of the development environment is usually smaller than the actual running environment. However, we should at least ensure the same configuration within the node, so as to avoid affecting the development progress.

Additionally, the software environment (compiler, debugger) should also be the same on the development and running platforms. In order to make full use of the heterogeneous hardware computing power, we often need to use a high-performance math library for heterogeneous hardware, and heterogeneous hardware tuning tools for program optimization.

4.2.3.1 Programming Language and Interface

(1) **General programming language and interface**

In order to improve the usability and compatibility, the development environments of heterogeneous system generally use the existing programming language and add their own specific extensions, instead of recreating a new programming language. Therefore, the general co-processor supports common languages such as C, C++ and Fortran.

In order to further facilitate the users, some organizations develop a general programming interface that enables programmers to use the same code to control different heterogeneous devices. OpenCL (Open Computing Language) is the most widely used heterogeneous programming interface.

OpenCL is the first free, open-source standard programming environment for general-purpose parallel programming of heterogeneous systems. It is also a unified programming environment to facilitate software developers in efficiently and easily coding for high-performance computing servers, desktop computing systems, and handheld devices. It is widely used in multi-core CPU, GPU, cell type architectures, DSP (Digital Signal Processor), and other parallel processors, and has big potential for future development on games, entertainment, scientific research, and medical applications. OpenCL consists of a language (based on C99) for writing kernels (functions operating on OpenCL device) and a set of API, which is used to define and control the hardware platform. OpenCL provides a parallel computing mechanism based on the task segmentation and data partition.

OpenCL supports GPU and MIC co-processors, and therefore can be commonly used among heterogeneous clusters. Programmers can use a unified code, ignore the hardware details, and make full use of the heterogeneous hardware accelerators.

(2) **MIC programming language and interface**

MIC supports C, C++, and Fortran, and extends these languages in a similar way to the OpenMP style. Similarly, MIC specific interfaces can take advantage of the hardware resources on the MIC card and improve the efficiency of the program.

4.2.3.2 Compiler and Debugger

We generally use the Intel Compiler XE Compiler Collection (including C/C++, Fortran, and other languages compilers) to compile MIC programs. Note that we need to choose the compiler version that supports MIC (i.e. version 2013 and later). Intel has partially opened the source of MIC drivers, operating system, and so forth, so that a free third-party compiler might emerge in the near future. If the user is extremely sensitive to the cost of the software, they can select the free version of the compiler in the future.

While the compilation is also divided into the CPU part and the MIC part, Intel has integrated these two versions into a same package so that programmers can invoke them in one go.

IDB is a debugger from Intel. IDB supports the debugging using a command line interface as well as JAVA-supported graphical interface. Therefore it is relatively easier to learn and use than GDB. IDB is fully compatible with GDB commands. IDB supports MIC kernel debugging, and is the preferred debugger for MIC programs.

4.2.3.3 High-Performance Math Library

MIC-based high-performance math function library provides high-performance math calculation for MIC. Due to the relationship between MIC and Intel CPU, Intel MKL math library can also be used on MIC.

Intel MKL math library includes linear algebra math library (BLAS, LAPACK, ScaLAPACK, Sparse Solvers, FFT) and vector math library (power function, trigonometric, exponential, hyperbolic, logarithmic function, random number generator, etc.).

As for MIC, high-performance math library can be manually called. We can set when to use the math library in the source code, and can also directly use these functions in MIC functions; we can also set the environment variable in the system, so that the case with a large number of computations can use the MIC-version functions, and in the case with a small number of computations we can still use the CPU-version functions. Or we can even make CPU and MIC work cooperatively by presetting the environment variables that specify the workload proportions of CPU and MIC.

As MIC is based on the popular x86 architecture we expect that there will be more and more high-performance math libraries developed for MIC, bringing more options for users.

4.2.3.4 High-Performance Tuning Tools

Since MIC is based on the x86 architecture and the Linux system, the vast majority of Intel performance tuning tools can be applied to MIC. Intel VTune Performance Analyzer is commercial software. We can automatically obtain most information of the programs by simply rerunning the program without recompiling. In addition to collecting the time consumptions of functions, VTune can also sample at the code-line level, and even at the assembler-statement level. VTune can sample different hardware events, so we can conduct more targeted analysis and testing. The latest version of VTune fully supports the hardware event sampling of MIC, so it can be well applied in the performance tuning of MIC programs.

4.2.4 *Further Reading*

- **The Intel Developer Forum** (IDF) is a technical workshop organized by the Intel Corporation. It is held twice a year in spring and autumn, in seven different regions, including the USA and China. IDF contains keynotes, technical seminars, and technical presentations. Keynotes are generally given by Intel's senior experts, with topics on the up-to-date technology developments. As a leading company on CPUs and network processor, IDF is indeed the best event for learning about the latest trends. More detailed information can be found from the following website:
  http://www.intel.com/content/www/us/en/library/viewmore.results.html?prTag=rauthorship:inteldeveloperforum
- **The MIC computing forum** is the most efficient and the most up-to-date platform for discussions on MIC technologies. You can obtain the assistance from the experts in the many various fields.
  https://software.intel.com/en-us/forums/intel-many-integrated-core
- More resources can be acquired from the following website:
  https://software.intel.com/en-us/mic-developer.

4.3 The Application Environment of CPU + GPU Heterogeneous Parallel Systems

4.3.1 *The Hardware Environment*

There is no difference between GPU and MIC in the system hardware structure, please refer to the previous related introduction.

4.3.2 *The Software Environment*

4.3.2.1 The Specific Software Environment for GPU

The specific software environment for GPU is different depending on the manufacturers. However, it generally requires hardware drivers, the runtime library, and other basic parts. In this chapter, we take CUDA as an example to discuss the related issues.

The software environment of CUDA is divided into the driver and runtime library. The driver's role is to generate links between software and hardware to ensure that the operating system can successfully utilize the hardware device. The driver is closely related to specific hardware. Therefore, we should download the exact version that corresponds to the hardware configuration. Note that the latest driver does not necessarily provide the best performance. The latest drivers are generally optimized for latest devices, and the performance of previous generation devices might degrade. In general, we should use a relatively new and official version of the driver. Although some beta-version drive might bring performance benefits, we generally do not use them for industrial applications due to stability considerations.

A runtime library contains a collection of libraries that need to be used when running the GPU programs. These libraries call the interfaces provided by the driver to perform GPU-based computations.

It should be specially noted that we need to configure the driver on each node that includes the GPU cards. For the runtime library, it is only needed by the application. Therefore, we can install one copy on a shared directory for all nodes. This approach can reduce the workload related to installation and maintenance. However, we should also pay attention to the control of written permission on the file under the shared directory to prevent unintended deletion of the file and to avoid a global software failure.

4.3.3 *The Development Environment*

We have described the application environment of the GPU-based heterogeneous parallel systems through the previous two chapters. This chapter provides a general introduction to the development environment, and intends to give readers a guide when choosing language, utilities, and libraries in the GPU program development.

4.3.3.1 Language and Libraries

Since 2007, GPU has been widely adopted in many different industries. Different applications bring different demands on metrics that range from software accuracy, speed, to the length of development cycles. Moreover, due to the diverse backgrounds, different programmers' capabilities of mastering a new programming language vary widely. In order to meet the various needs of the market, NVIDIA and its partners have developed a wide variety of programming methods, such as CUDA C/C++, CUDA Fortran, OpenACC, HMPP, CUDA-x86, OpenCL, JCuda, PyCUDA, Direct Compute, Matlab, and Microsoft C++ AMP. This large set of tools makes it possible for programmers to find the right programming tool that suits their project and programming habits.

Multi-threaded programming languages can be divided into two categories. One is the compilation directive type, that is, by adding the corresponding compilation directives before a section of serial code, the compiler can automatically compile the serial code into parallel programs. The other option is the explicit multi-threading model. The user can directly write parallel programs, and call GPU threads explicitly. In contrast, the compilation directive is relatively simple, easy for beginners to learn, and generally leads to a shorter development cycle. However, as the parallelization is automatically controlled by the compiler, the operation on GPU is relatively inflexible.

In the following programming languages that we introduce, OpenACC and HMPP are the compilation directive languages, and are similar to OpenMP for CPUs. The CUDA Fortran compiler supports both compilation directives and an explicit multi-threading model. Other languages only support the explicit multi-threading model. Among these, CUDA-x86 is a special tool that can directly translate the CUDA C/C++ program into CPU multi-threaded programs, without manual modification of the code. Table 4.1 shows the detailed information about the various languages.

In order to support multiple programming languages, and to improve the developer's efficiency, NVIDIA and its corporation partners, collaborated with developers in various domains, to make some of the most commonly used algorithms into modules, and to build a group of powerful and efficient libraries that cover a wide range of application domains. Developers can directly call these libraries without redevelopment by themselves.

Table 4.1 GPU programming languages

Programming language	Supported language	Developer	Classification
CUDA C/C++	C/C++	NVIDIA	Explicit multi-threading model
CUDA Fortran	Fortran	PGI	Explicit multi-threading model/compilation directive
OpenACC	C/C++ Fortran	PGI/CAPS/GREY/NVIDIA	Compilation directive
HMPP	C/C++ Fortran	CAPS	Compilation directive
CUDA-x86	CUDA C	PGI	Explicit multi-threading model
OpenCL	C/C++	Khronos	Explicit multi-threading model
Direct Compute	C/C++ integrated in Direct3D	Microsoft	Explicit multi-threading model
Microsoft C++ AMP	C/C++	Microsoft	Explicit multi-threading model
Jcuda	JAVA	Personal	Explicit multi-threading model
PyCUDA	Python	Personal	Explicit multi-threading model

(a) AmgX provides a simple solution to accelerate the core part of a linear solver. For computation-intensive linear solvers, AmgX can gain a speedup of 10×, and is ideal for the implicit non-structural methods.
(b) cuDNN is the GPU acceleration library designed for deep neural networks, and can be integrated into a high-level machine learning framework.
(c) cuFFT implements the GPU-based Fast Fourier transform (FFT), and provides a simple interface for developers to use.
(d) cuBLAS-XT is based on cuBLAS and provides multi-GPU acceleration support for Level 3 BLAS. Developers do not need to configure and schedule the multi-GPU resources.
(e) NPP is a powerful package that includes thousands of image processing and signal processing functions.
(f) CULA Tools is a linear algebra library provided by EM Photonics. It provides GPU acceleration for complex mathematical calculations.
(g) MAGMA supports the latest standard LAPACK and BLAS linear algebra libraries, and provides special optimization for heterogeneous computing.
(h) IMSL Fortran Numerical Library is developed by RogueWave. It is a GPU-based mathematical and statistical library, covering a wide range of contents.
(i) cuRAND provides high-speed GPU-accelerated random number generators.
(j) ArrayFire is a content-rich GPU library, covering the fields of mathematical signal processing, image processing, and statistics. It supports C/C++, Java, R, and Fortran.
(k) cuBLAS is a GPU-accelerated BLAS library, can gain 6× to 17× acceleration compared to MKL BLAS.
(l) cuSPARSE provides acceleration for sparse matrix computations.
(m) Thrust provides an open-source library for high-speed parallel algorithms, with GPU-accelerated sorting, scanning, conversion and other commonly used algorithms.
(n) NVBIO is for the biological domain, provides GPU acceleration for high-throughput sequence analysis and short- and long-read alignment.

In addition to the libraries mentioned above, there are still many other libraries, such as the Triton Ocean SDK for real-time simulation of ocean water, NVIDIA VIDEO CODEC SDK for GPU hard and soft decoding, HiPLAR, which is a linear math library based on R, OpenCV, which is a widely used library in computer vision, a computational geometry engine GPP, and a sparse matrix iterative solver, Paralution.

4.3.3.2 The Development Tools

The development tools can generally be grouped into three categories: compiler tools, debugging tools, and performance analysis tools. The following describes the debugging tools and performance analysis tools. Finally, Nsight, an integrated development environment, is introduced.

(1) **The debugging tools**

Cuda-Gdb is a debugging tool that is released together with NVIDIA CUDA SDK. It is based on Gdb command line, and supports CPU and GPU debugging. Developers with previous Gdb experience will be able to get started quickly.

Together with CUDA-GDB, CUDA-MEMCHECK can quickly locate the memory access errors of GPU functions, and settle the "unspecified launch failure" error in the procedures.

Allinea DDT supports GPU clusters and single nodes, and supports MPI, OpenMP and CUDA program debugging.

TotalView is a GUI-based debugging tool that can support multi-process and multi-thread MPI, OpenMP and GPU debugging.

(2) **The performance analysis tools**

All CUDA performance analysis tools are developed based on CUDA Profiling Tools Interface (CUPTI), with different focuses on the program and the analysis targets that they support. Among these, NVIDIA Visual Profiler can provide visual feedback and basic optimization recommendations when the developers optimize CUDA C/C++ programs. It supports Linux, Mac OS X, and Windows platforms. Now it also supports remote single-node analysis.

NVPROF is a CUDA C/C++ analysis tool under a command-line interface and the results can be sent to the Visual Profiler display. It allows flexible functionality through user-written script.

In addition, a number of commercial softwares, such as TAU Performance System, VampirTrace, and the PAPI CUDA Component, can support CUDA, pyCUDA, and HMPP hybrid parallel programming. They contain much richer features than Visual Profiler, and support multi-node analysis. They can help developers have a better overall picture of the program and cluster situation.

(3) **Nsight**

Nsight is an integrated development environment developed by NVIDIA. It integrates compiling, debugging, and performance analysis functions. There are two versions, the Nsight Visual Studio Edition on Windows platform, and the Nsight Eclipse Edition on Mac OS and Linux.

Code Development Nsight provides updated development templates, and integrates CUDA SDK samples, so that developers can quickly get started and write CUDA applications. It also provides code highlighting, auto-completion, and navigation of functions.

Debugging Nsight is able to debug CPU and GPU at the same time in a similar way. It supports remote debugging, and contains the CUDA-MEMCHECK module for memory error detection.

Analysis Nsight provides an interface and functions similar to Visual Profiler, supports various performance data analysis, and can automatically generate optimization suggestions. Moreover, Nsight supports hot-spots analysis in the Eclipse version, and can automatically tag hot statements in the GPU functions.

4.3.4 Further Reading

- **CUDA ZONE** provides a one-stop guide and resources download for GPU application development, covering all aspects of the information from the entry, development tools, to the online learning resources. The website also contains real-time updates on the latest news in the field of GPU computing, and is suitable for all application developers using GPU. https://developer.nvidia.com/cuda-zone
- **GTC** The annual GPU Technology Conference is the largest and most important GPU computing event for technology exchange and discussions. Scientists, engineers, and researchers from different fields all around the world give research talks and exchange successful experiences. Participants can learn about the latest industry trends and research results. The talks are made into online videos, which can be accessed from the following website:
http://on-demand-gtc.gputechconf.com/gtcnew/on-demand-gtc.php
- The **GPU computing forum** is the most efficient interactive platform. You can obtain the assistance from the experts in the many various fields. See more resources via https://devtalk.nvidia.com/default/board/53/gpu-computing/
- **Nsight**'s developing the environment can be downloaded from the website:
http://www.nvidia.com/object/nsight.html.

Chapter 5
Supercomputer System Performance Evaluation Methods

5.1 Introduction to HPC System Performance Evaluation

5.1.1 The Significance and Challenge of System Performance Evaluation

For high-performance computers or supercomputers, the powerful computing performance is the most important value and advantage. How to evaluate the performance of a high-performance computer system is of general concerns to users, manufacturers and research institutions. Performance evaluation is also an indispensable step in the development, selection, introduction and effective utilization of a computer. The answers to the following frequently asked questions are all more or less related to the system performance evaluation:

"Does the designed computer system meet the design goals?"

"What is the gap between the performance of the designed system and the requirement of the application, and how can the current situation be improved?"

"How much performance improvement can hardware upgrades (CPU, memory, coprocessors or other components) bring?"

"What kind of the high-performance computer architecture is most suitable for this particular application?"

"Is it necessary to equip the high-performance computer with accelerator components?"

"What is the performance difference between vendor A's products and similar products manufactured by vendor B?"

"With a fixed budget, how could we get the best performance?"

... ...

The performance of a computer system is the joint result of the hardware, the software, and many other factors. The hardware part includes the processor, the

ASC Community, *The Student Supercomputer Challenge Guide*,
https://doi.org/10.1007/978-981-10-3731-3_5

memory, the storage devices, the network equipment, the system bus and the recently popular coprocessor (accelerators). The software part mainly includes the operating systems, the compilers, the programming languages, the parallel communication libraries, the math libraries and other systems software. Since the evaluation is concerned with the performance status of the system itself, the software typically does not include the actual application software. Compared with an ordinary PC, a workstation or a server, a high-performance computer system is much larger and more complex in both the hardware and the software parts. It is a difficult challenge to accurately evaluate the performance and give answers to the above questions.

From an architectural perspective, the cluster system is the mainstream due to its excellent performance cost ratio. In addition, the traditional massively parallel processing computer (MPP) system still keeps its position due to its irreplaceable architectural and performance advantages. The products of Cray, IBM, Fujitsu and other mainstream MPP manufacturers can be further divided into SMP, cc-NUMA, vector machines and other different types. Computers with different architectures have inherent differences in the performance term.

For computing components, the current mainstream CPU mainly includes Intel/AMD x86-64 processors, IBM's Power processors, Fujitsu SPARC 64 processors and ARM processors. Even for the same type of processors (such as Intel x86 64-bit processors), different versions demonstrate differences in the chip architecture, the manufacturing technology, the number of cores, the cache and the instruction set, resulting in the differences in their performances. In addition to CPU, in recent years, NVIDIA GPU, Intel Xeon Phi and other coprocessors also have been widely adopted, which in turn greatly increases the complexity of the performance evaluation.

From the perspective of the system software, operating systems, compilers, communication libraries (including the underlying communication protocol and the upper library) and core math libraries all have significant impacts on the performance. For example, the Linux operating system becomes the first choice for high-performance computers because it can be customized and simplified, while there is still a small number of computing systems that use other operating systems such as Unix and Windows, and an even smaller number of virtual-machine-based high-performance computing systems. Different operating systems have different performances (although usually insignificant). Also, the core math library BLAS has many versions, for example, GotoBLAS, OpenBLAS, ATLAS, MKL, and they have their own advantages in the performance of different systems.

Every computer system ultimately serves a certain application or software, and different software applications typically have different performance features; even for the same software, the performance will also be different because of the different hardware, system software, operating parameters, or run time environments.

The complexity shown above makes it difficult to find a unified set of performance evaluation rules or metrics to review all the different computers. Accordingly, system performance evaluation approaches, methods and software demonstrate a large diversity, and a quick development with the dynamic change of computer hardware and software technology and the change of various users' needs. In recent years, for example, due to the growing size of high-performance computers (up to several million cores), the system's power, stability and reliability have become the hot issues. This requires that performance evaluation should not only consider computing performance, but also consider the power needed to obtain the performance and the duration for which the system can stably output that performance.

High-performance computers, or supercomputers, have been known as the pearl of crown in the computer architecture field. On one hand, they have superior performance and are extremely expensive; on the other hand, high-performance computing systems typically reflect the most advanced computing technology, and are far larger and more complex than the ordinary computer system. This not only makes the performance evaluation extremely important, but also brings great challenges to the evaluation work.

5.1.2 The Metrics and Main Contents of System Performance Evaluation

The performance of computer systems typically covers the computing performance, the storage (I/O) performance and the communication performance. For high-performance computers the most important is undoubtedly the computing performance.

Since the vast majority of high-performance computers are used to deal with floating-point calculation, the most commonly used performance measure is floating operations per second (FLOPS). The floating-point operation refers to the addition, subtraction, multiplication and other basic operations.

For a given high-performance computer system, its theoretical peak performance is fixed. In a typical cluster system, for example, the system theoretical peak performance is calculated as (CPU main frequency) × (floating-point operands per clock cycle per core) × (total number of CPU cores).

Consider the following configuration of a cluster system,

- Processor Type: Intel Xeon E5 2692 V2@2.2 GHz, 12 cores
- Number of processors per node: 2
- Total number of nodes in the system: 16

A single-core Intel Xeon E5 processor can execute 8 floating-point calculations per clock cycle, then the theoretical peak performance of this cluster system is: $2.2 \times 8 \times (12 \times 2 \times 16) = 6758.4$ GFLOPS = 6.7584 TFLOPS.

Here, TFLOPS is the abbreviation for Tera-FLOPS (Note: from Kilo, Mega, Giga, Tera, Peta to Exa, each increase is 1000 times larger than the previous level, e.g., 1000 Kilo = 1 Mega, 1000 Mega = 1 Giga, and so on). Currently the ordinary high-performance computers already provide a computing performance in the scale of tens or hundreds of TFLOPS, and a few of the most advanced systems can reach the level of PFLOPS. The current most powerful system in the world is the Tianhe-2 system located at the national supercomputing center of Guangzhou in China, and the theoretical peak performance reaches 54.9 PFLOPS. Nowadays, the exascale computer is the common goal of manufacturers and research institutions worldwide.

When a computer system is equipped with the GPU, Intel Xeon Phi or other accelerators, the theoretical peak performance is the sum of the CPU peak performance and the accelerator peak performance.

From the above formulas, the theoretical peak performance only relies on the CPU and accelerating parts (coprocessors), and does not involve other hardware such as memory, network or I/O devices. Obviously, this is inconsistent with what we usually experience with the practical system performance: the system with a large memory and fast network will, typically, be more powerful than that with a small memory and slow network. Because of this, the theoretical peak performance is just the performance upper boundary that the computer system cannot exceed. The actual performance is influenced by other factors, and can only be obtained through evaluation.

The performance evaluation obtains the performance parameters of the system by running the specific evaluation software. In addition to the main computational performance (FLOPS) mentioned above, many other items are also included, such as the system bandwidth, network latency, memory bandwidth and other individual performance metrics. Users can also customize the metrics (such as the number of tasks completed in one unit time). No matter what metric is adopted, the application is always the most important criterion.

The vast majority of evaluations will have a simple metric. For some metrics, bigger is better, such as the bandwidth, the memory capacity and the storage capacity. In contrast, for some other indicators, smaller is better, such as the running time, memory usage and the latency. A good evaluation usually has the following characteristics:

- Easy to run: without strict requirements on either the user, the equipment, or the environment;
- Reproducible: this is the one of the basic requirements, otherwise the evaluation does not make much sense;
- Easy to understand: the results can be easily accessible, and can easy to understand (e.g. output a single value to indicate FLOPS);
- Representative: it can represent a wide range of practical application scenarios, or is very close to the actual performance of the system;
- Scalable: the test can be performed on different scales of parallel systems;
- Public: there is no restriction about showing the results to a public audience;
- Transparent: open source would be best;
- Appropriate run time: the evaluation time should not be too long.

According to the test contents, test metrics and test methods, the performance evaluation of high performance computers can usually be divided into the following four categories:

(1) **Single-metric evaluation**

It is also called a micro-benchmark. The evaluation is usually testing a specific component of the system or an individual performance metric, such as the memory bandwidth test, the network performance test or the I/O performance test. Its main purpose is to test whether the system component, or the performance metric, has met the design goals, to perform troubleshooting and to compare different components horizontally (such as evaluating different network devices).

(2) **Overall computing performance evaluation**

The overall performance evaluation of the computer reflects the performance in general situations. Common tests include HPL, HPCC, SPEC HPC and so on. The overall evaluation is usually closely associated with theoretical peak, but may have a large difference from the theoretical peak. The overall performance is the result of the realistic runs of specific evaluation software, and can reflect the performance of the basic characteristics of the system. At present, most domestic and international supercomputer rankings are based on the overall computing performance evaluation results.

(3) **Domain-specific application performance evaluation**

As different applications usually demonstrate quite different performances on the same computer system, the overall computing performance evaluation results are often not consistent with the actual performance of a specific application, and can be largely different in some cases. However, when we look at applications in the same domain, which are generally based on a common set of computing models and methods, domain-specific evaluation software can usually provide a good estimation of the application performance in the corresponding domain. NPB from NASA and the IAPCM Benchmarks from Beijing Institute of Applied Physics and Computational Mathematics in China are typical examples of domain-specific performance evaluation software.

(4) **Performance evaluation for a typical application**

As all the high-performance computing systems are designed to run one or a certain number of application software, people's ultimate performance evaluation is usually towards the goal of achieving the best computing performance of the target application software. Therefore, for users with specific target software, the best choice for performance evaluation is to build the evaluation software based on the target application software. As mentioned earlier, a good evaluation needs to meet certain conditions, and simply running the target software several times is far from a good performance evaluation. A lot of effort is still needed for converting practical application software to a typical performance evaluation benchmark. For example, we still need to prepare standard test data sets, tune the suitable parameters of the application software, develop methods for checking the accuracy of the results and design output formats that are easy to read and compare.

5.1.3 The Procedures and Methods of System Performance Evaluation

Although there is no consistent standard or procedure for the system performance evaluation, the evaluation process can usually be divided into three steps: pre-testing preparation; performance testing; and post-testing result collection and analysis.

5.1.3.1 Pre-testing Preparation

The pre-testing preparation mainly includes system preparation, recording of the configurations and making testing plans.

The system preparation includes the setup of the hardware and software system, the testing site and the running environment, which are required steps for assuring a smooth run of the testing process. Recording of the configurations refers to the process of writing down the details of the hardware/software configurations for making the tests, which would generally include the following items:

A. Hardware

- Node type and number of nodes;
- Network: manufacturer, model, main configurations, bandwidth and latency; and
- Node information:
  - Main BIOS configurations;
  - CPU: number, type, frequency and stepping;
  - Memory: type, capacity (the total number, and the capacity of each one), speed and latency;
  - Hard disk: number, type and capacity;
  - Accelerator: model and number;
  - Chassis power: model and power; and
  - Chassis fan: number, position and power.

B. Software

- Operating system: model, version and main configurations;
- Compiling system: model, version and main compiler parameters;
- Testing software: version, running workloads and number of processors used; and

Table 5.1 Parameter settings of IOMETER testing plan

Access mode	Block size	IOPS	Bandwidth	Latency	CPU utilization
Sequential Read	4K				
	16K				
	64K				
	512K				
Sequential Write	4K				
	16K				
	64K				
	512K				
Random Read	As above, breakdown each line according to the different block size				
Random Write					
Mixing Random Read and Write					

- Open source software: version of the compiler, compiler options and the version of the software library that it uses.

Making testing plans refers to designing the workflow of the testing process, and designing tables or other formats of files for recording the results. The parameter setup is shown in Table 5.1 for the storage performance test of IOMETER. When testing, once an item is completed, we fill the corresponding slot in the table, which keeps the testing process in order, and a good record of the entire procedure.

5.1.3.2 Performance Testing

When running the testing software the most important thing is to ensure the stability and reliability of the testing process. Usually we follow the following principles:

- The minimum system principle: shut down unneeded services, non-essential programs, and the graphical interface.
- The exclusion principle: run the test program only, without interference from other programs and users.
- The no-interference principle: perform no other operations other than the testing itself.
- The repetition principle: repeat the testing process multiple times and choose the (best) stable results.
- Single-adjustment principle: modify only one parameter at a time, when the modification of the testing parameters is needed.

5.1.3.3 Post-testing Result Collection and Analysis

After collecting all the test results, we firstly need to confirm the reliability of the results, that is, the differences between different runs of the same configuration should be within the allowable error range (typically <5%). After confirming the results, we need to process the test results, and demonstrate the results in the form of charts or tables.

Finally, you can analyze the results, to see whether the expected objectives are met; if not, we need to find out the reasons and to make improvements.

5.2 HPC Systems Performance Evaluation Methods

In the previous session, we mentioned the four major evaluation methods for a typical application: the single-metric evaluation; the overall computing performance evaluation; the domain-specific application performance evaluation; and the performance evaluation. These four evaluation methods are currently the most commonly used methods and each of them focuses on different parts in practical evaluation tasks.

This chapter introduces the main testing methods and the corresponding benchmarks. For the single-metric evaluation part, we introduce the memory performance benchmark STREAM and the network communication evaluation program OMB; for the overall computing performance evaluation part, we introduce HPL and HPCC; for the domain-specific application performance evaluation, we introduce NPB, IAPCM and GRAPH 500; for the performance evaluation of a typical application, the scenario would be completely different for different applications, which we do not discuss here.

The ranking of high-performance computing systems based on the results of the performance evaluations has long been a hot topic in the HPC community. At the end of this chapter, we will introduce the major ranking lists, such as the TOP500, TOP100 in China and GREEN 500.

5.2.1 The Main Evaluation Programs and Applications

5.2.1.1 Memory Performance Evaluation Program STREAM

With the increase in the number of processor cores, memory bandwidth is more and more important for the improvement of the overall system performance. If the system is not able to quickly transfer the data from the memory to the processor, a number of processing cores will wait for the memory data transfer and become idle. The idling time not only reduces the efficiency of the system but also offsets the performance improvement that multi-core and high frequency bring. The STREAM benchmark developed by University of Virginia is currently the most popular memory bandwidth test tool.

STREAM has two language versions, C and Fortran. STREAM also provides an MPI version, the results of which are uniformly measured as MB/s, reflecting the sustained memory bandwidth of the system.

STREAM is simple to run. For the serial version, we just need to compile and run on the target platform directly. For the parallel version (multi-threaded, OpenMP or MPI), you only need to compile and run in a similar way to common parallel programs (for example, the MPI version requires an installation of the MPI library in the system, the file "stream_mpi.f" needs to be compiled using mpif77, and we need to run the program using mpirun).

Note that, although the overall performance evaluation benchmark set HPCC uses the MPI version of STREAM to measure the memory bandwidth, in practical applications, the memory access performance largely depends on the communication network. Therefore, the memory bandwidth evaluation on the stand-alone system becomes more important. The serial and multi-threaded versions (including pthread and OpenMP) are more frequently used.

5.2.1.2 Network Communication Evaluation Program OMB

As the cluster system with multiple nodes interconnected by a network is the current mainstream form of high-performance computers, the communication performance is critical for high-performance computers. OSU Micro-Benchmarks (OMB) from Ohio State University is one of the industry's most well-known communication performance testing tools.

OMB currently supports three communication models (MPI, UPC and OpenSHMEM), and the latest version also provides support for CUDA and OpenACC. One of the most typical and most common use cases is undoubtedly the MPI communication performance tests.

OMB provides a rich set of testing scenarios, including point-to-point communication, collective communication and unilateral communication. Each type of communication also provides multiple outputs, such as latency, bandwidth, multi-threaded latency and multi-threaded bandwidth. Different communication data sizes can be set (such as latency tests using different message sizes [1 B, 4 B, 16 B, 64 B, and 256 B], and bandwidth tests using different message sizes [4 KB, 64 KB, 256 KB, 1 MB, and 4 MB]).

OMB uses a common GNU build system (configure and make) and performs the tests using operating parameters such as osu_latency and osu_bw.

5.2.1.3 Floating-Point Performance Evaluation Program HPL

High Performance LINPACK Benchmark (HPL) is currently the most well-known high-performance computer performance benchmark software, and the evaluation results of HPL have become a basis for supercomputers' performance ranking of the world's TOP500 list and the TOP100 list in China.

LINPACK (LINear algebra PACKage) is the function library for solving linear algebra problems (solving equations, least squares, etc.), written by Jack Dongarra and colleagues in the 1970s and the 1980s. To help users estimate the time needed to solve equations using LINPACK, LINPACK developers wrote the first version of the LINPACK Benchmark, and provided the test data sets: according to a problem with the 100×100 matrix, they gave the test results (run time) conducted in many popular computer systems (a total of 23 kinds) at that time. In this way, users can infer the time required to solve the problem of their own matrix. In June 1993, HPL results were introduced in the global TOP500 list of supercomputers' performance, and replaced the original peak performance ranking as the new basis. This new ranking basis was widely accepted and received quickly. Thereafter until today, HPL has been the most important high-performance computing performance evaluation program.

HPL uses the time taken for solving a dense linear equations $Ax = \mathrm{b}$ to evaluate the computer's floating-point performance. In order to ensure the fairness of the evaluation results, HPL does not allow the user to modify the basic algorithm (using Gaussian elimination of the LU decomposition) that is, the user must ensure the same total number of floating-point calculations. For the $N \times N$ matrix A, the total number of floating-point calculations for solving $Ax = \mathrm{b}$ is $(2/3 \times N^3-2 \times N^2)$. Therefore, as long as the size of the problem N is given, and the system computation time T is measured, we can compute the HPL floating-point performance as $(2/3 \times N^3-2 \times N^2)/T$, with the unit of FLOPS.

HPL allows the user to select any N scale, and modifying the algorithm and program without changing the total number of floating-point operations and computation accuracy. This prompts the users to put all their efforts into obtaining a better value of HPL performance.

Common HPL optimization strategies or approaches include:

(1) Select an N as large as possible. Before the system runs out of memory, the larger N, the higher the HPL performance. (Why is that? Think about the reasons.)
(2) The key computation of HPL is the matrix multiplication (accounting for more than 90% of the total time), and the matrix multiplication is implemented by using the block matrix multiplication algorithm. The size of the block has a huge impact on the computing performance. To choose the best block size, you need to carefully consider the CPU cache size and other factors, and perform small-scale tests to identify the best block size.
(3) HPL uses MPI for parallel computing. In the procedures, computing processes are distributed in a two-dimensional grid, you need to arrange the processor array and set the grid size. Similarly, you may want to perform small-scale tests to get the best solution.

(4) LU decomposition parameters, MPI, BLAS math libraries, compiler options, operating systems and many other factors also have an impact on the final test results. Readers can refer to the relevant literature to know more details.

HPL's installation requires a compiler, MPI parallel environment and basic linear algebra libraries (BLAS). It should be noted about the choice of BLAS library. As mentioned before, the current commonly used BLAS library has multiple versions (GOTO, OpenBLAS, Atlas, MKL, ACML, etc.), the performance of different implementations on different systems may be quite different. Please refer to the relevant literature or need of the actual test to choose the best version.

5.2.1.4 A Comprehensive Performance Evaluation Program HPCC

The High Performance Computing Challenge (HPCC) is a HPC system evaluation benchmark set that is released and supported by DARPA's High Productivity Computing System (HPCS) project.

HPCC consists of several well-known test programs, including a number of single-metric evaluation and floating-point performance evaluations. It deliberately chooses typical evaluation programs that have either distinct spatial locality or temporal locality, and expects to give a full performance evaluation of a high-performance computer system. The seven typical evaluation programs that HPCC selects are as follows:

- HPL, as previously described, can obtain the floating-point performance—both spatial locality and temporal locality are good;
- DGEMM, double-precision floating-point number matrix multiplication, can obtain the floating-point performance—both spatial locality and temporal locality are good;
- STREAM, as previously described, can obtain the sustained system memory bandwidth performance—spatial locality is good while temporal locality is poor;
- PTRANS (parallel matrix transposition) reflects system communication performance—spatial locality is good, while temporal locality is poor;
- RandomAccess tests random memory update rate (GUPS)—both spatial locality and temporal locality are poor;
- FFT, double precision complex one-dimensional discrete Fourier transform, obtains floating-point performance—spatial locality is poor, but temporal locality is good;
- b_eff measures the system communication bandwidth and latency.

These seven evaluation programs can obtain a total of eight test values (b_eff provides two results, bandwidth and latency), and HPCC often adopts a visual image shown in Fig. 5.1 to exhibit a computer's performance (different colors of lines represent different computers).

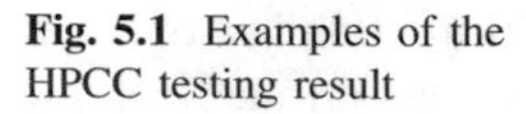

Fig. 5.1 Examples of the HPCC testing result

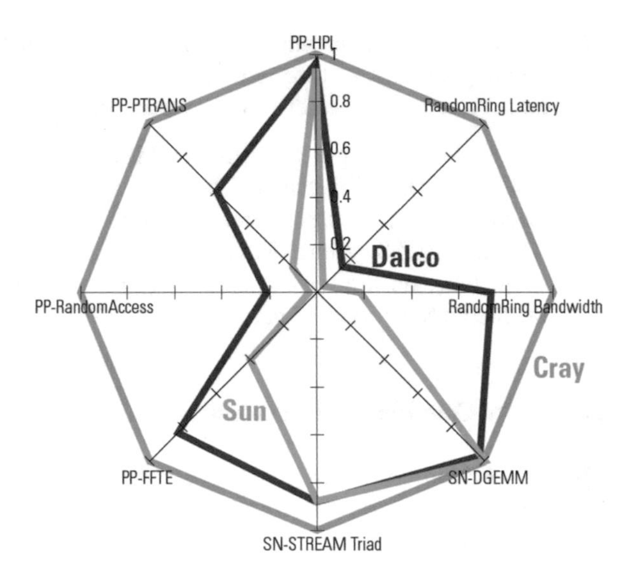

Although HPCC offers extensive test results that are far more than a single performance evaluation program (such as HPL), it is not widely supported or accepted in the high-performance computing industry. The reasons are various. Among these, that test procedures and test results are too complex, and that a single index for comparison is not easy to produce, are the two undoubtedly important factors. There was a time that worldwide researchers attempted to integrate these eight indicators into one single index just like FLOPS of HPL. However, despite of some minor bright spots, this work ultimately failed. Nevertheless, HPCC is still an excellent benchmark for performing a comprehensive assessment of high-performance computer system performance.

5.2.1.5 The Field Evaluation Program Set NPB

NAS Parallel Benchmark (NPB) is a parallel computer performance evaluation benchmark in the scientific computing domain. NPB contains eight different benchmarks, all from computational fluid dynamics applications, and each benchmark simulates a different kind of behavior of the parallel application. Therefore, NPB can evaluate the CFD parallel application

NASA Ames Research Center has developed NPB, and set the standard for NPB. Within the standard range, the manufacturers can implement their own program in an optimal way according to the specific characteristics of different machines.

For a parallel version of the NPB, you need to do a specific optimization on parallel granularity, data structures, communication mechanisms, processor mapping, memory allocation and other aspects, according to the architecture of the system. The versions of standards above NPB 2 provide a unified parallel program with MPI implementations.

NPB contains five core programs, and they are frequently applied algorithms:

(1) Integer Sort (IS) mainly tests integer performance and collective communication performance. IS's main operation is parallel sorting, which is important for the program code of particle physics research. IS does not contain floating-point arithmetic, but it uses a great amount of data communication. IS is sensitive to communication latency. For such application, the use of high-bandwidth, low-latency communication network is critical for improving the performance.
(2) Embarrassingly Parallel (EP) has almost no data communication, and mainly tests floating-point performance of mathematical functions. Communication network has little effect on the EP result. Due to the low dependence on the communication network performance, EP generally has better scalability.
(3) 3-D Multigrid (MG) adopts a multi-grid algorithm for solving three-dimensional Poisson equations, and requires that the number of processors must be a power of two. MG mainly tests structured non-contiguous memory access collective communication and point-to-point communication. MG has both integer and floating-point arithmetic.
(4) Conjugate Gradient (CG) mainly tests unstructured collective communication and point-to-point communication.
(5) Fast Fourier Transform (FFT) uses the 3D FFT to solve partial differential equations, and mainly tests collective communication.

NPB also includes three linear solvers, which all come from computational fluid dynamics:

(1) Lower-Upper Triangular (LU) uses symmetrical over relaxation (SSOR) to solve block sparse equations and it requires the number of processors to be a power of two. LU mainly tests fine-grained non-contiguous memory access point-to-point blocking communication, and is very sensitive to the communication latency.
(2) Block Tridiagonal (BT) solves 5×5 block tridiagonal equations, and requires that the processor grid must be square. BT mainly tests the balance between communication and computation. The communication mainly contains non-continuous memory access point-to-point long-message communication.
(3) Scalar Pentadiagonal (SP) solves five-diagonal solution equations, and requires that the processor grid must be square. SP mainly tests the balance between communication and computation, and mainly contains non-continuous memory access point-to-point long-message communication. The communication mode is quite similar to that of BT, but with a quite different communication/computation ratio. The communication strength of SP is higher than that of BT.

Each benchmark of NPB has six different Classes, namely A, B, C, D, W (Workstation) and S (Sample), except that IS has no Class D. A, B, C and D represent four different scales of the problem, and the problem scale increases from

A to D. The problem scale refers to the size of the data set used for testing. The larger data set requires a longer computation time and larger storage space.

The number of processors used in the test also needs to be specified. NPB's eight programs have different requirements on the number of processors. BT and SP require that the number of processors must be n^2 (n is a positive integer), LU, MG, CG, FT and IS require it to be $2n$ (n is a positive integer), while EP has no special requirement on the number of processors. If the specified number of processors cannot meet the requirements there will be a corresponding error message at compile time.

For more accurate system performance, in the test, you need to choose the proper scale of the problem and the proper number of processors according to the system configuration and research requirements.

5.2.1.6 Domain-Specific Application Performance Benchmark Set: IAPCM Benchmarks

Similar to the NPB in the USA, the IAPCM Benchmarks reflect the requirements of a specific field. IAPCM Benchmarks are developed by IAPCM, one of the China's main research institutes that develop high-performance computing applications.

IAPCM Benchmarks have been developed since the early 1980s, and was first published in the 1992 27th issue of Computer World with a paper named "Performance evaluation for a number of computer systems". Currently, there are ten programs in the benchmark set:

(1) The EPT program for the dynamics of non-equilibrium systems: the program uses a six-point difference scheme to approximate the non-equilibrium thermodynamic photon diffusion equation (E equation), and adopts GEAR integral to approximate bound electron occupation probability equations (Pn equation) and free electron temperature equation (T equation). This is an extremely complicated unstable problem. Particularly, the Pn equation is a set of non-linear strong stiff ordinary differential equations (rigidity ratio up to 10^{10}).
(2) Two-dimensional tracking interface activity grid program (YGX): YGX is a two-dimensional computation program that combines the Euler method and the Lagrange method. The differential discretization is based on axisymmetric hydrodynamic integral equations. This method is used to track the movements at the boundaries. In the interior region, this method uses a set of similar-to-elliptic equations to build the computing grid, so the grids can be further divided into finer grids continuously.
(3) Particle transport program (DOSE): DOSE adopts the Monte Carlo method to solve the particle transport equation.
(4) Linear muffin potential orbital method program (LMTO): LMTO adopts the linear energy band theory method to implement the program, which is currently one of the most widely used methods in the world.

(5) Iterative solver I for linear algebraic equations (WBR): after discretization, a heat conduction equation will generate a set of linear algebraic equations, and its coefficient matrix has five non-zero diagonal elements. The WBR program adopts the block over-relaxation iterative method (BSOR) to solve linear algebraic equations.
(6) Iterative solver II for linear algebraic equations (KGFM): after discretization, heat conduction equations with the first derivative terms generate a set of linear algebraic equations, and its coefficient matrix has five, seven or nine non-zero diagonal elements. The KGFM program adopts incomplete LU decomposition preconditioned conjugate gradient method (ILUGG) to solve linear algebraic equations.
(7) Two-dimensional discrete ordinate method program (FSW): FSW solves the two-dimensional neutron transport equation. The angular velocity space adopts the discrete ordinate method, and the geometric space uses difference approximation on the rectangular grid. The GS iterative method is used as the solver.
(8) Two-dimensional multi-layer implicit Euler method program (NNFC): NNFC uses the Euler-type numerical method. The discretization is based on axisymmetric hydrodynamic differential equations. This method uses a fixed Eulerian grid, and different substances are marked with different labels on the grid. It uses the movement of labels between grid points to calculate the amount of the transport.
(9) Molecular dynamics program (MD): The MD program adopts Morse hot to describe intermolecular interactions, and uses the staggered grid. The program design is based on the linked-list data structure.
(10) State equation programs (DUP): This is part of the database program of the state equations, and its function is to pre-process the data of the state equations for the interpolation afterwards. This program has a lot of data accesses, and uses multiple channels for data transfer. So DUP can be used to evaluate the performance of the data transfer channels.

The test results of the above benchmark programs provide the necessary information for the development and purchase of computer systems, which can also be used as a reference for institutes that perform similar modeling experiments.

5.2.1.7 Domain-Specific Application Performance Benchmark Set: GRAPH 500 Benchmark

GRAPH 500 is a ranking list for high-performance computing systems' performance on data-intensive applications. The corresponding evaluation program set is the GRAPH 500 benchmark set. While most of the above benchmarks we discussed focus on the floating-point computation performance and the traditional scientific computing domain, GRAPH 500 mainly uses graph theory to analyze and rank the supercomputers' throughput for simulating biological, security, social and other complex problems.

The problem that GRAPH 500 benchmark processes is a breadth-first search in a huge undirected graph. The benchmark includes two computing kernels: the first is to generate the graph for searching, compress and store the graph in sparse CSR or CSC formats; the second is to search using parallel BFS methods. There are six different data sizes available:

- Toy scale, 226 vertices, about 17 GB memory
- Mini scale, 229 vertices, about 137 GB memory
- Small scale, 232 vertices, about 1.1 TB memory
- Medium scale, 236 vertices, about 17.6 TB memory
- Large scale, 239 vertices, about 140 TB memory
- Huge scale, 242 vertices, about 1.1 PB memory

GRAPH 500 uses GTEPS (109 Traversed edges per second) to rank the systems. Currently, the top one system on the list is the Japanese K supercomputer, with the performance of 17977.1 GTEPS.

5.2.1.8 Floating-Point Performance Evaluation Program HPCG

High Performance Conjugate Gradient (HPCG) is a new evaluation program recently proposed by Jack Dongarra, one of the main contributors of HPL. The goal is to evaluate certain metrics that are not covered by HPL. It uses the conjugate gradient method to solve large-scale sparse matrix equations $Ax = b$.

In fact, this type of equations is from the unsteady nonlinear partial differential equations. In the iterative solving process, it requires frequent and irregular accesses to the data. Therefore, HPCG requires high bandwidth, low latency and high CPU frequency of the computer system. Such systems are generally costly, and require a long cycle to develop. Even in the United States, only a few leading supercomputers belong to this type.

Overall, the applications that HPCG covers are still a small portion in the entire HPC domain. To make the HPCG benchmark as widely acceptable as HPL, there is still a long way to go.

5.2.2 High-Performance Computing System Rankings

5.2.2.1 The TOP500 List

TOP500 is the most famous and influential ranking list of the supercomputer systems. The first ranking list was released in 1986 by Professor Hans Meuer of Germany, which was based on the theoretical peak of the system. After the introduction of HPL as the new ranking metric in 1993, TOP500 list has been recognized as the most important list of supercomputer systems in the world. TOP500 is currently released twice a year, one at the International Supercomputing Conference held in Germany in June, and the other one at the Supercomputing Conference held in the United States in November.

5.2.2.2 The TOP100 List in China

The high-performance computer performance TOP100 list (HPC TOP100) in China is the list of the fastest 100 computer systems based on the LINPACK performance. It provides an important indicator of the development of China's HPC systems.

In 2001, to promote the development of HPC, and to promote the installation and industrialization of the domestic high-performance computers, the Specialty Association of Mathematical and Scientific Software (SAMSS) of the China Software Industry Association proposed the ranking list of Chinese high-performance computing systems. In 2002, Jiachang Sun, Guoxing Yuan and Linbo Zhang released the first TOP50 list. After more than ten years of development, the TOP100 ranking list has been recognized and accepted by the Chinese high-performance computer development institutes, system vendors, and high-performance computing applications institutes, and other relevant research institutions, and has an extensive and far-reaching influence home and abroad. The TOP100 list does not only reflect the development of the HPC industry in China, but also reflect the country's rapid development on technology and economy.

Both the international TOP500 list and China's TOP100 list choose HPL as the performance metric for ranking, mainly because HPL has the following features:

- Good demonstration of requirements for a wide range of applications. HPL requires high floating-point computing performance and large memory of the system. The requirements are quite common for most major HPC applications, such as development of military equipment, energy, aerospace, aviation, high-end equipment manufacturing, biomedicine, petroleum, marine environmental engineering, weather/climate forecasting, etc.
- Strong adaptability. HPL is adaptable to a wide range of different systems with different architectural features, such as different processors, different memory architectures, different accelerators, different network, different operating system, etc. For example, the China's HPC TOP100 list includes clusters, MPP, the world's first GPU system, the top one system in the world with Intel Xeon Phi accelerators, and Windows clusters.
- Good scalability. Since 2002, in China's HPC TOP100 list, the performance increases by six orders of magnitude (from 13.17 GFLOPS to 33.86 PFLOPS).
- Good evaluation of the system reliability. HPL is one of the most effective programs that can detect whether the system can be running stably and providing correct results. The evaluation of reliability is particularly important for the current large-scale systems with tens of thousands of computing

components. Reliability is one of the biggest challenges for designing a large-scale system. The first step for running HPL is to ensure the stability of the system.

5.2.2.3 The GREEN 500 List

GREEN 500 re-ranks the TOP500 systems based on their energy efficiencies. The evaluation metric for the ranking is the computing performance achieved for each watt, measured in FLOPS/W.

There has been so far 15 announcements of the GREEN 500 list since 2007. The current top one system is the Tokyo Institute of Technology's TSUBAME-KFC-LX 1U-4GPU/104Re-1G cluster system, with the performance of 4389.82MFLOPS/W.

The power consumption of a computer system can be measured through a specific power meter. Note that the power consumption value should be the average system power consumption during a stable stage (as shown in Fig. 5.2).

5.2.3 Further Reading

- **Memory performance evaluation program STREAM** The latest version is available from the website of http://www.cs.virginia.edu/stream/. This website also gives the current TOP20 systems evaluated by STREAM. The currently top one system is the large shared memory Altix UV2000 computer from SGI.
- For **OMB** download links and more detailed information, see http://mvapich.cse.ohio-state.edu/benchmarks/.
- **HPL** The latest version of can be obtained from http://www.netlib.org/benchmark/hpl/.

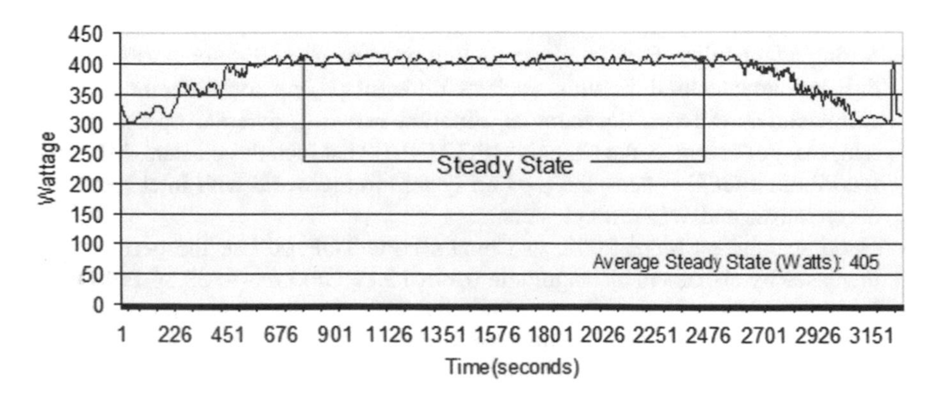

Fig. 5.2 The record of the measured system power consumptions

- **HPCC** can be obtained from http://icl.cs.utk.edu/hpcc/.
- **NPB** can be obtained from http://www.nas.nasa.gov/publications/npb.html.
- **TOP500** Detailed information about TOP500 can be found from http://www.top500.org.
- **China's HPC TOP100** More information can be found from http://www.hpctop100.cn.

5.3 HPC Application Analysis and Monitoring Tools

Supercomputers are built to improve the efficiency of scientific research development and industrial/agricultural production. Therefore, one design goal is to have an increasing computing speed. However, the supercomputer is a very complex system, which includes the underlying hardware platform, system middleware platforms and also the various applications running on the supercomputer. How to maximize the efficiency of different applications on the supercomputer system is an extremely complicated problem. It involves the optimization of hardware, system and software. In addition, differences in the application domains lead to huge differences in the characteristics of the application software. How to identify the major characteristics of a given application and to optimize accordingly is the key of the problem. The process involves three steps: the extraction of profiling data; the analysis of the extracted data; and the analysis of the characteristics of the application, which will be discussed.

5.3.1 The Extraction of Application Characteristic Data

Typically, the extraction and analysis of the application characteristics is based on a specific hardware platform and a specific computing scenario. When the hardware platform or the computing scenario changes, the demonstrated application characteristics will often change correspondingly. Therefore, we firstly determine the specific hardware platform and the specific software configurations, and then perform the collection of the application data. Table 5.2 lists the hardware platforms for common tests. For the configuration of the platform, we generally choose the most common parameters and the mainstream CPUs. When choosing CPU, we have to consider the CPU frequency, QPI, cache and so on. When choosing memory, we need to consider the memory capacity, the number of memory channels and so forth. For the IO system, we need to consider the disk read/write speed. For the network system, network bandwidth, latency and so on should be considered. After the platform is determined, we need to decide on the test cases and compile the application software. After we compile the application software, we need to test the application to make sure that the specific case is running

Table 5.2 Hardware platforms for common tests

Component	Configuration	Related considerations
Workload	Running time	Scale and configuration of the test cases
Software	Version	Versions of compilers and application software
Hardware type	Compute node: NX560T (blade)	Choose the hardware type according to the project requirements
CPU	Intel E5-2650, X5675	CPU frequency, cache, QPI frequency, Turbo configuration, NUMA configuration
Memory	48/64 GB DDR3 1333 MHz	Memory capability, memory frequency, DIMM/ socket
Storage/file system	IO node, NFS	AS500G3
Network	InfiniBand	InfiniBand, 10G ethernet
OS	RHEL6.1	OS version

normally, and ensure the effectiveness of the collected application characteristics data.

After the above preparations, we can start the characteristic collection work, which consists of three different parts: system level, application level and micro-architecture level. On the system and application levels, we focus on the computing platform's CPU, memory, network and disk utilizations; while on the micro-architecture level, we need to consider the instruction execution efficiency, CPI, vectorization efficiency and so on. After the basic characteristic collection we can perform the optimization of the system platform and the application software, and run the characteristic collection iteratively until there is no apparent bottleneck. In the end, we analyze the application characteristics according to the collected data, and provide the basis for the optimization of the system design. Figure 5.3 shows the general workflow of the HPC application analysis system. The detailed metrics of the performance tests and the monitoring tools are listed in Table 5.3.

During the application execution, we use the tools listed in Table 5.3 to monitor all the different metrics, including the CPU usage ratio, average load, system occupancy rate, floating-point computing performance, CPI, vectorization efficiency, the memory capacity requirement, memory bandwidth usage, the size of input/output file, output frequency, output pattern, network flow, network bandwidth latency, the size of data blocks, the times of major message transfers, the proportion of MPI communication in the whole application and so forth.

In order to identify accurately the characteristics of the application software, the time interval of the characteristic extraction is often very short, and usually measured in seconds, which results in a huge amount of collected data. Therefore, efficient collection and analysis of the characteristic data often requires a database, such as Mysql, to complete.

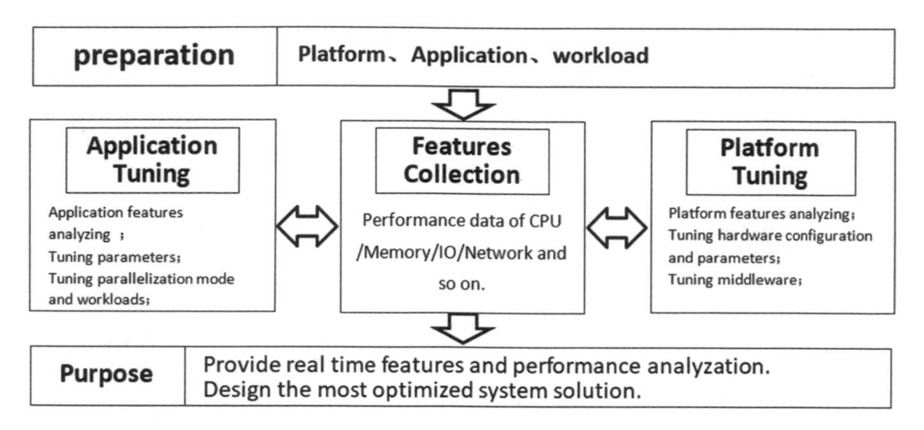

Fig. 5.3 The general workflow of the HPC application characteristic analysis system

Table 5.3 HPC performance metrics and monitoring tools

Major metrics		Description	Monitoring tools
CPU	Application computing time	Application Computing Time	ITAC, TAU
	Average load	CPU load (an indicator about whether more computing resources are needed)	TOP, ganglia, Teye
	CPU utilization rate	The ratio of effective CPU utilization by the application software and the system	TOP, PAPI, TAU, Teye
	L3 cache miss rate	A relatively high L3 cache miss rate suggests the bad efficiency of program data prefetching	PAPI, Vtune
	CPI	The average clock time consumed by each instruction, a relatively high CPI suggests a low efficiency and a large potential for further optimization	Vtune, Teye
	Floating-point performance	The real-time floating-point performance when running the application	Teye
	Vectorization efficiency	The ratio of the vectorized instructions in the total floating-point instructions	Teye
Memory	Memory capacity	The memory space consumed by the application	Free, NMON, Ganglia, Teye
	Memory bandwidth	The effective memory bandwidth consumed by the application	Teye
	SWAP usage	Swap space usage	NMON, Ganglia
	NUMA ratio	The proportion of remote memory accesses	Vtune

(continued)

Table 5.3 (continued)

Major metrics		Description	Monitoring tools
Network	Communication time %	same as MPI Time%, the proportion of MPI communication time in the entire run time	ITAC, TAU
	MPI wait time %	The proportion of the waiting time in MPI communication	ITAC, TAU
	Network in	Network in	Nmon, Ganglia, ethtools, Teye
	Network out	Network out	Nmon, Ganglia, ethtools, Teye
	Latency	The inter-node latency when running the application	OSU, IMB
	Bandwidth	The inter-node bandwidth when running the application	OSU, Iperf
I/O	Disk IO %	The proportion of IO time in the total run time	Nmon, Ganglia, iostat
	File system IO bandwidth	File system IO bandwidth when running the application	IOzone, Iometer
	Granularity	The size of the each data access when running the application	Nmon, Ganglia, Teye
	Frequency	The frequency of data accesses when running the application	Nmon, Ganglia

5.3.2 *The Analysis of Application Characteristics—Experimental Induction*

After the extraction of the application characteristic data, we need to perform the analysis. In general, for existing computers using the Von Neumann architecture, the analysis is primarily about four aspects: processor, memory, I/O and network.

5.3.2.1 Memory Characteristic Analysis

Memory is considered as the warehouse that connects to the CPU via a bridge. The capacity of memory determines the size of the "warehouse", and the bandwidth of the memory determines the width of the "bridge"—these are generally called "memory capacity" and "memory speed". Some HPC applications need a larger memory capacity (such as 256/512 GB) to run because of the special requirements of the algorithm or the modeling scenario. Some other applications need higher memory bandwidth (memory speed) to keep the application running efficiently. Therefore, memory plays an important role in HPC.

Depending on the different requirements for memory, HPC applications can be divided into two categories: memory-bandwidth demanding applications and memory-capacity demanding applications. Memory-bandwidth demanding applications refer to the applications that perform frequent data transfers between the CPU and the memory controller while the program is running. The memory bandwidth is generally the performance bottleneck in the application, as shown in Fig. 5.4. As can be seen from Fig. 5.4 the running application software has basically exhausted the maximum memory bandwidth resources that the system platform can provide. In this case, the memory bandwidth is the biggest bottleneck of this application software.

Thus, in terms of the memory use of a given application, we should analyze both the memory capacity and the memory bandwidth. We can obtain memory capacity usage from the algorithm that the application software uses, and we can also monitor it by using NMON, Teye and other tools. For the memory bandwidth usage, the sensitivity can be judged by adopting different processes combinations within the node, or be real-time monitored by some tools, such as Teye.

5.3.2.2 I/O Characteristic Analysis

Storage is an indispensable part of an HPC system. It is often the most technology demanding and the most difficult design component of the supercomputer, and usually the most expensive part as well. The storage performance has an essential impact on the application performance, especially for these data-intensive industries.

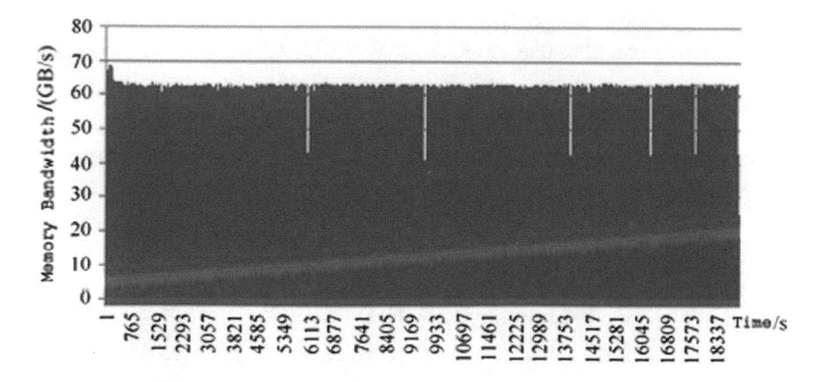

Fig. 5.4 The real-time memory bandwidth of running a specific application (DDR3 1333 MHz)

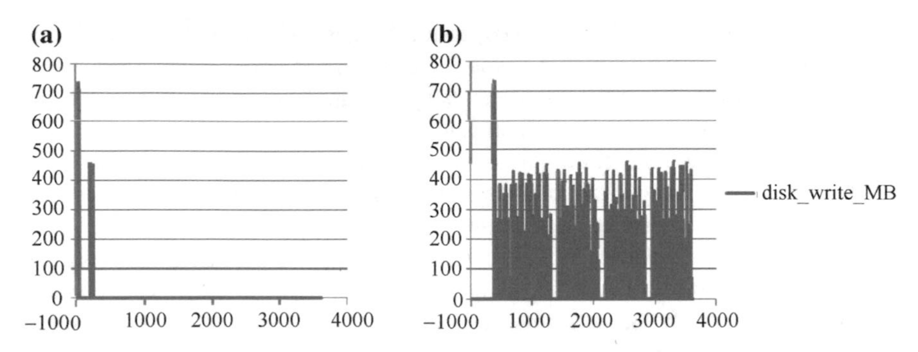

Fig. 5.5 The real-time I/O characteristics of running a specific application, **a** infrequent I/O read/write **b** frequent I/O read/write

In some application domains, such as bioinformatics, weather/climate modeling and computational fluid dynamics, the scale of the original input data can be as large as several TBs, and the output file after computation can be even bigger. Therefore, intuitively, the capacity of the storage system is an important factor that we have to consider in high-performance computing. In addition, some applications will generate huge intermediate and temporary data files, and these files have various types of formats and different sizes, so you need the high IOPS and high read/write rates of storage to improve the efficiency of the application. We give the I/O characteristics of two different running applications software in Fig. 5.5. The real-time I/O of the application in Fig. 5.4a is infrequent; while large and frequent I/O read/write are required in the application software shown in Fig. 5.4b.

5.3.2.3 Processor Characteristic Analysis

During the execution of the application, CPU spends a large amount of time carrying out arithmetic computation and evaluation of logic expressions. Therefore, the CPU usage is usually high, as shown in Fig. 5.6. We can use the TOP command to check roughly the CPU usage, the occupancy of the system, the process waiting time and so on. We can also use tools, such as PAPI, TAU, VTUNE, NMON, Teye

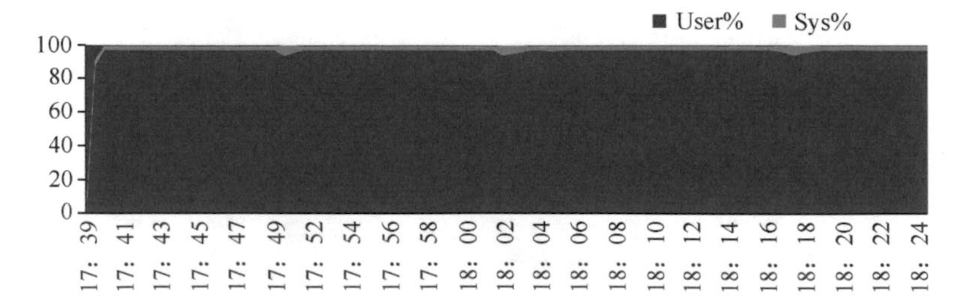

Fig. 5.6 The real-time CPU usage ratio and the running time (s) of a specific application

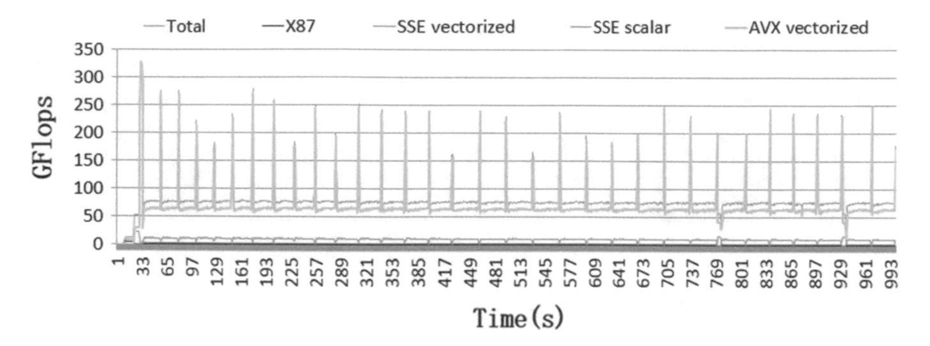

Fig. 5.7 The real-time floating-point operation speed and the running time (s) of a specific application

and so forth to monitor the detailed CPU characteristics during the execution of the application, and to perform a deep analysis. Figure 5.7 shows the real-time floating-point operation speed during the execution of the application. CPU usage is highly important in application performance analysis, so reasonable and accurate analysis methods and measurements are the keys of analysis.

When we analyze the CPU requirements for different applications, we can monitor the CPU usage ratio, floating-point operation speed, vectorization ratio and other indicators to understand the degree of the optimization. These can guide us to adjust the application performance parameters or hardware platform parameters to obtain the optimal performance, so as to achieve the goal of the application optimization.

5.3.2.4 Network Characteristic Analysis

The modern supercomputers usually have hundreds of thousands or even hundreds of millions of CPU cores. If the application software wants to take full advantage of these computing resources, there is inevitably communication and data exchange between the processors. As for the interconnection and data exchange between the processes, there are process-based, thread-based as well as shared-memory-based communication modes. The most widely used is the message passing mode, namely MPI. Therefore, by monitoring MPI's communication time proportion and network flow during the application operation it is possible to analyze the network characteristics of the application, as shown in Fig. 5.8. These data can be obtained from Ganglia, ITAC and other tools.

In addition, sometimes you would like to deeply optimize MPI-based applications, particularly at the code optimization level. You can monitor and extract MPI message events and the usage of the MPI functions, such as MPI_Allgather, MPI_Alltoall, MPI_Scatter and other functions. In the case of large-scale parallel

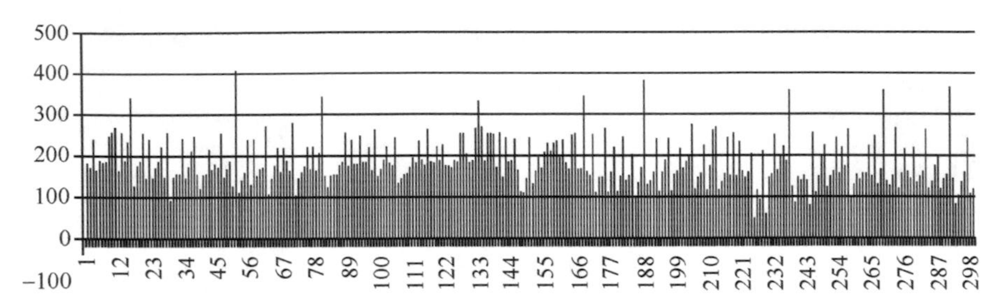

Fig. 5.8 The real-time network speed and the running time (s) of a specific application

computing these functions have enormous impacts on the performance of applications. Such data can be obtained from VTune, TAU and other tools.

The methodology of the above characteristics analysis is an experimental, routine, and standardized procedure. You can apply this method to the analysis of any software, and summarize its application characteristics that provide the necessary guidance for the system configuration, cluster designing, and the application optimization. However, the conclusion of this analysis is merely a summary of the demonstrated program behavior, and often cannot identify the underlying causes of software-related characteristics. A full understanding of an application and its performance requires a comprehensive set of both the basic domain knowledge and the algorithm knowledge of the application software. The following chapter will give a more detailed discussion on this part.

5.3.3 The Analysis of Application Characteristics—Theoretical Predictions

In the previous chapter, we proposed a set of application characteristic analysis method for HPC application software. By recording the resource occupancy ratio or other characteristics of the application running on a particular platform, we can evaluate the performance of the application. On the one hand, the analysis provides suggestions and feedback to the design of a specific hardware platform. On the other hand, the analysis also provides the basis for further improving the application software. Although the results acquired by the profiling method are precise, the disadvantages and limitations are also evident. Firstly, the recorded indicators are only the reflectors of the characteristics of the specific tested platform, which is platform-specific. To acquire accurate analysis results we would need to do testing on many platforms, generating a huge workload. Secondly, as mentioned earlier, since HPC involves a wide range of industries and a huge amount of application software, it is impossible and impractical to profile and analyze the characteristics of all the different application software. Moreover, the analysis results acquired from profiling are all superficial phenomenon of the application software. Without understanding the underlying mechanism it is hard to give a reasonable explanation of the different performances of different applications. So, do we have a better solution to solve the above problem?

Based on the authors' years of HPC research and practice experience we are able to establish a set of theoretical performance prediction methods. Different from the profiling method, it is independent from the specific hardware platform, and uses the basic physics principle, the model, and the algorithm to analyze the performance characteristics of the application software.

We know that no matter how big the application software is, how many functions it includes, how friendly the UI, it is all implemented by program codes. The codes are built based on specific physical principles, proper physics models and appropriate mathematical algorithms. Figure 5.9 demonstrates this dependency.

The theoretic model of a software application includes the basic physics model and the corresponding numerical algorithm. The physics model generally refers to the physics framework or mechanism that the application software tries to simulate. For example, for material and chemistry application software, we generally use the first-principle method (such as Hartree-Fock method, self-consistent field iteration); for biomolecule or medicine design applications, we may use molecular dynamics theories and methods; for industrial designs, we may use finite element methods; for electromagnetism or optics applications, we may use Maxwell equations; for weather applications, we may use thermodynamic equations, state equations. These basic physical principles and methods determine the algorithm and the parallel model we can use in this application. Therefore, the key issue that the HPC application software needs to handle is how to solve the above physics problems. The mathematical algorithms are the basic mathematical solutions used to solve the above physics equations. Similarly, different mathematical algorithms lead to different performances and different levels of parallelism of the application software. If we can perform a clear theoretical analysis of the resource utilization for solving the

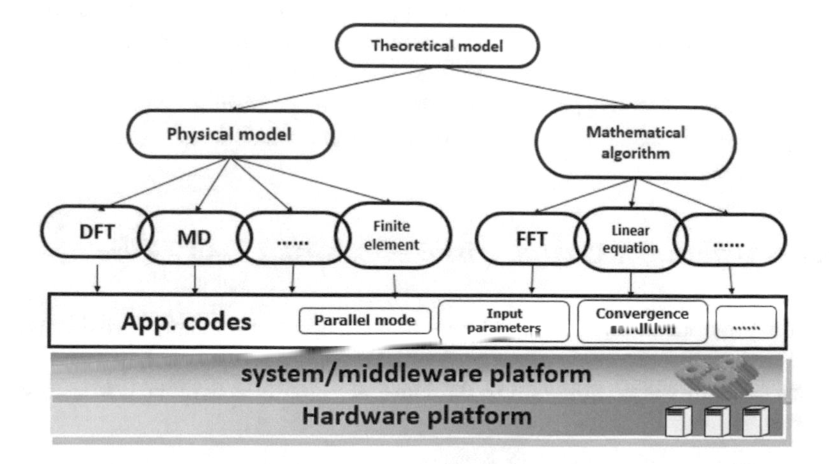

Fig. 5.9 The theoretical performance prediction method

related physics problems using corresponding mathematical methods, we can thoroughly understand the characteristics of any application software without profiling.

To give a detailed description, we take an application example from the material and quantum chemistry field. As mentioned earlier, material and quantum chemistry are mostly simulating the problem using the first principle in quantum mechanics, and calculates the specific physics variables using the self-consistent solving of Kohn-Sham equations. However, there are many different numerical methods for solving the Kohn-Sham equations. These different solving methods lead to different parallel designs that demonstrate completely different computation patterns. Table 5.4 lists some common basic methods used by the HPC application software in this domain. From Table 5.4 we can see that the basis vector of VASP is plane wave, and it uses the inter-band and inter-basis vector parallelism to perform parallel processing. Since the basis vector of plane wave is far away from the description of the real orbital functions, the computation requires the processing of a large number (hundreds of thousands) of plane waves. Besides, during the inter-band and inter-basis vector parallelization, each iteration requires the orthogonal normalization operations to all the basis vectors, which results in frequent memory read/write and inter-process data communication, thus leading to high memory bandwidth requirement and high network throughput. In contrast, though Abinit uses plane wave basis vectors, it uses inter-K-point mode to perform parallel processing. Each K point in the space is complete. Unlike inter-band or inter-basis vector parallelization, it does not need to transfer a huge number of normalization parameters between the processes. Therefore, we can see the feature of Abinit is high memory bandwidth with small network throughput, which is different from VASP.

We can see from the above analysis procedure that theoretical prediction method could rapidly identify the major characteristics of the application, which can bring important guidance for the optimization of the application and the design of the hardware platform. However, it is also apparent that the theoretical prediction method alone is far from enough, and cannot provide precise characteristic data. Therefore, the theoretical analysis method should be combined with the profiling method mentioned in the previous section, so as to perform a comprehensive and thorough analysis of the application software's characteristics.

5.3.4 Application Characteristic Monitoring Tool—Teye

From the previous discussions, we know that the application characteristic analysis work relies on tools to extract the characteristic data. We have already mentioned a number of open-source software tools. Here we highlight another powerful characteristic monitoring tool—Teye.

Table 5.4 The basic physics principles used in some material and quantization application software

Application software	Basis vector		Parallelization				Diagonalization				
	Plane wave	AO	Inter-band	Inter- basis- vector	Inter-K-point	Spin polarization	CG	PC-CG	DAV	RMM	CP
VASP	√		√	√					√	√	
Abinit	√				√	√	√	√			
PWSCF	√			√	√						√
Siesta		√									
Castep		√			√		√				
CPMD	√										√
GAUSSIAN		√									√

Teye stands for Teye high-performance application characteristics monitoring and analysis system. It is an excellent software developed by the Inspur's professional HPC team. Teye is mainly used for the extraction of system resource usage of the high-performance application that runs on the large-scale cluster systems, and reflects the run-time characteristics of the application. Teye helps users to maximize the computing potential of the current platform, and thus provides a scientific guidance to optimize the system, optimize the application and adjust the application algorithm.

Teye is a tool with GUI, consisting of two ports, one for cluster performance monitoring and extracting performance indicators, and the other for client-side performance analysis. It is based on BS architecture, and users can use Teye software through the web browser without any installation to monitor and extract the performance indicators. Figures 5.10 and 5.11 show the login interface and the main application interface respectively. In addition, Teye also has a number of

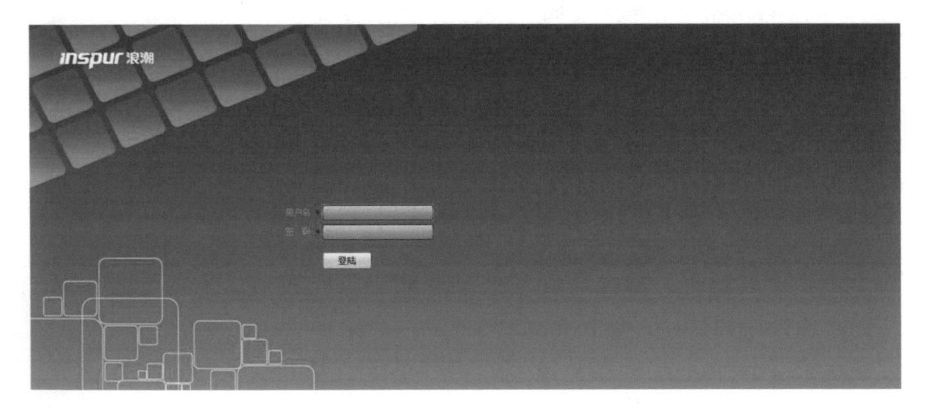

Fig. 5.10 The login interface of Teye for performance monitoring and indicator extraction

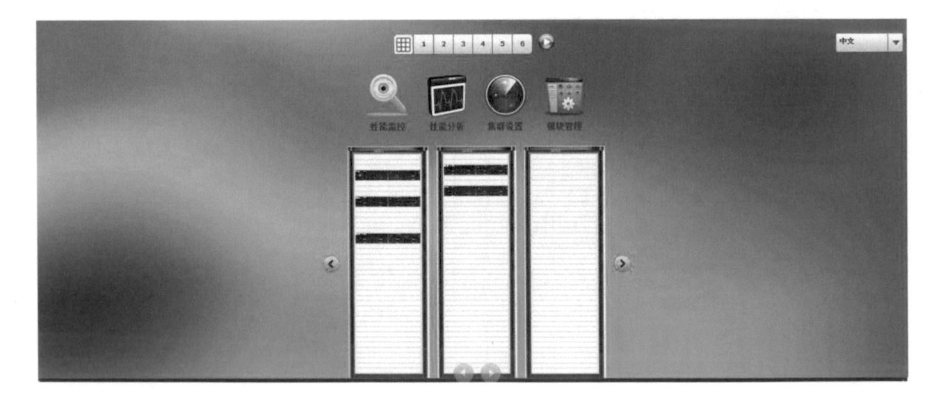

Fig. 5.11 The main application interface of Teye for performance monitoring and indicator extraction

advantages over other tools. For example, it is small in size, easy to use, accurate in run-time monitoring and has a low resource utilization. Even under heavy system load, Teye's demand for the system resources is far less than one thousandth, which greatly ensures the accuracy of the profiling results.

Teye is a quite powerful tool. It can monitor the characteristics data of processor, memory, network and IO, thus reflecting the run-time characteristics of the application software. Moreover, Teye can also grab more than 40 microarchitecture indicators, and nicely meet the needs of application tuning. These indicators include: inter-node CPU single and double floating-point performance, x87 floating-point unit performance, SSE/AVX unit performance, 128bit SSE instruction vectorization rate, 256bit AVX instruction vectorization rate, inter-node memory read/write bandwidth, intra-node NFS file system throughput, total NFS throughput and the data exchange bandwidth between PCI-E device and host memory.

Figure 5.12 shows the application interface of Teye during the monitoring and indicator extraction. It reflects the key performance indicators of the current focused nodes (specified by the user), which are demonstrated dynamically by sketches. Besides, we can simply click the node icon to obtain the real-time performance monitoring curves, shown in Fig. 5.13.

Moreover, in order to facilitate the management and use of monitoring data, Teye also provides support for the MySQL database so that the user can easily store and manage the monitoring data as a database. In addition, Teye also allows the monitoring data to be exported as the input of the analysis port for further performance analysis, as shown in Fig. 5.14.

To mine deeply the performance data, which can fully reflect the performance characteristics of the running application, Teye also provides a performance analyzing tool. This tool mainly analyzes the performance data exported from the port of performance monitoring and extraction. It provides a variety of approaches for

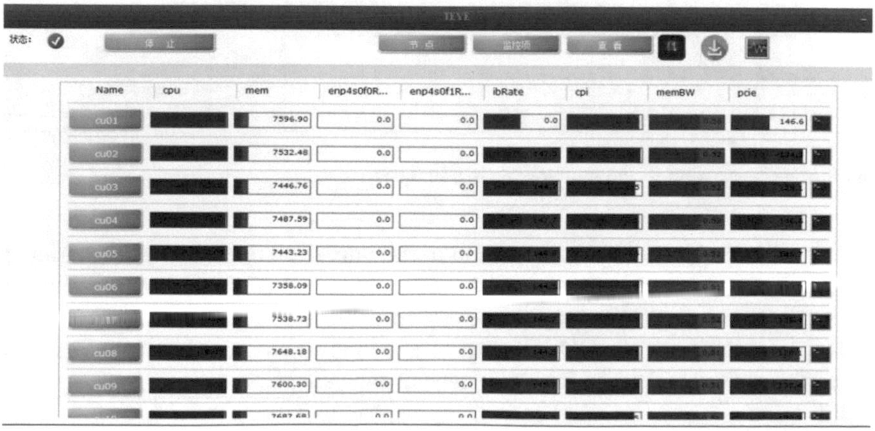

Fig. 5.12 Performance monitoring interface of Teye

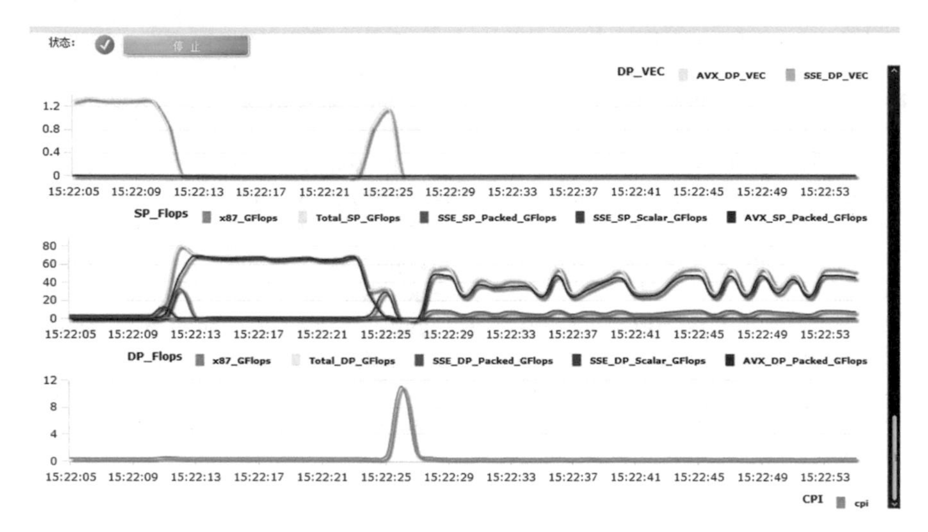

Fig. 5.13 Real-time performance curves of Teye

TEYE

状态: 启 动 节 点 0 15 30

名称	开始时间	结束时间	节点	状态	导出	删除	下载
vasp_SO	2015-10-	2015-10-	cu01,cu05,cu02,cu03,cu	finish			
roms_tse	2015-10-	2015-10-	cu01,cu05,cu02,cu03,cu	finish			
lmp_larg	2015-11-	2015-11-	cu01,cu05,cu02,cu03,cu	finish			
g1	2015-12-	2015-12-	cu03,cu04	finish			
opm1	2015-12-	2015-12-	cu01,cu05,cu02,cu03,cu	finish			
lmp_zhul	2016-01-	2016-01-	cu01	finish			
ctqmc	2016-01-	2016-01-	cu01,cu02,cu03,cu04	finish			

刷新

节点 全选

Fig. 5.14 Management of the monitoring data in Teye

statistics, filtering, locating hot spots and data comparison operations. In addition, it utilizes charts, performance radar, report generation and other various presentation forms, and makes it easier for users to export charts and reports.

Teye analysis port is easily installed in a Windows environment, and Fig. 5.15 shows the main interface of the performance analysis port.

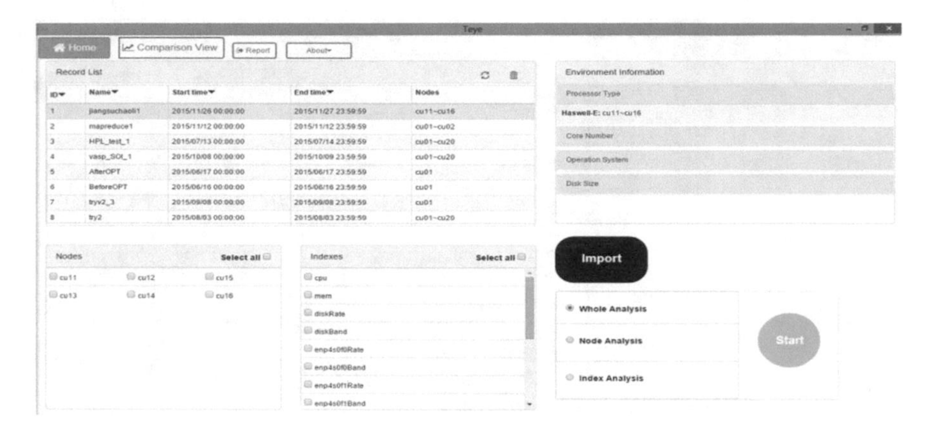

Fig. 5.15 The main interface of Teye performance analysis port

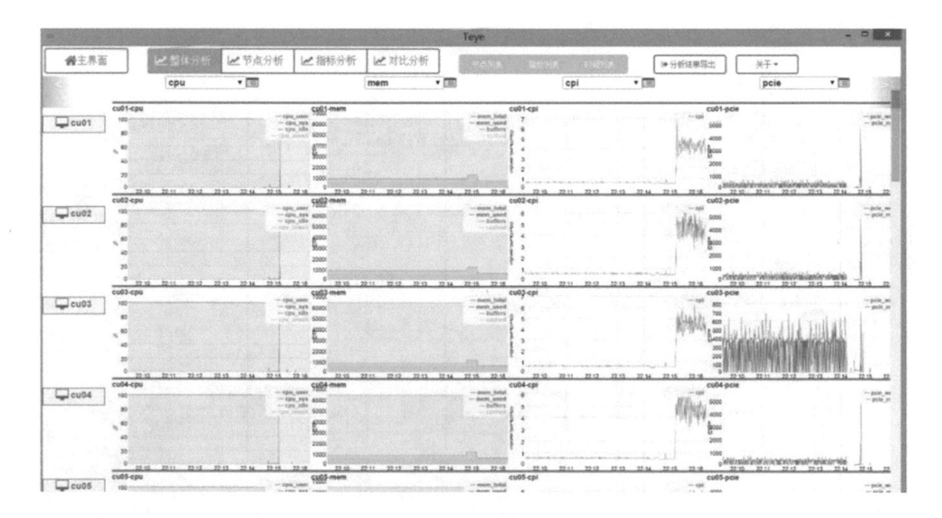

Fig. 5.16 The overall analysis interface of Teye analysis port

Figure 5.16 shows Teye's overall analysis function, Fig. 5.17 shows its node analysis function, Fig. 5.18 shows its indicator analysis function and Fig. 5.19 shows its comparison analysis function. The overall analysis is used for overall control and gives a comprehensive demonstration on the nodes and performance indicators for the monitored job. The node analysis can penetrate every compute node to analyze the running load and resource dependency, and generates the corresponding performance radar map. The indicator analysis is used for

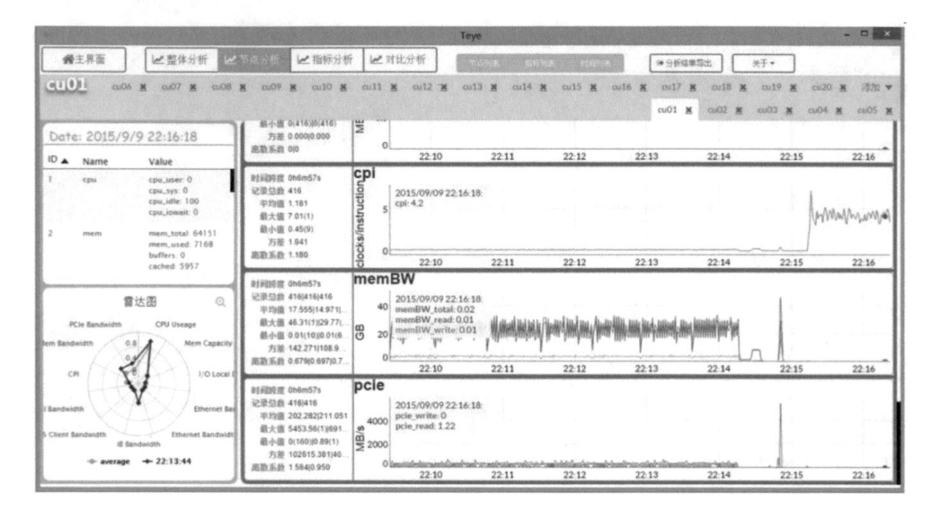

Fig. 5.17 The node analysis interface of Teye analysis port

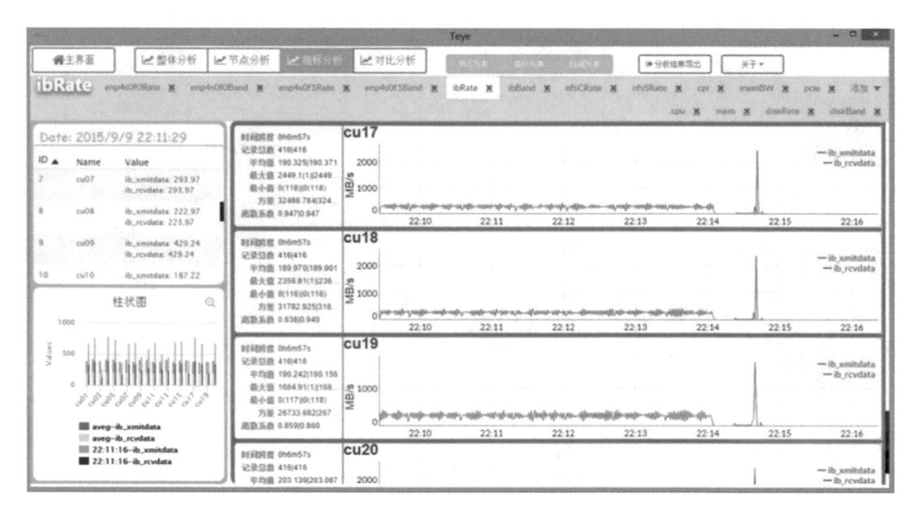

Fig. 5.18 The indicator analysis interface of Teye analysis port

comparison among different nodes with the same indicator, reflecting the load balance or process asynchronism of the application. The comparison analysis function allows the user to define the indicator parameter according to the needs, which greatly facilitates the performance analysis and optimization.

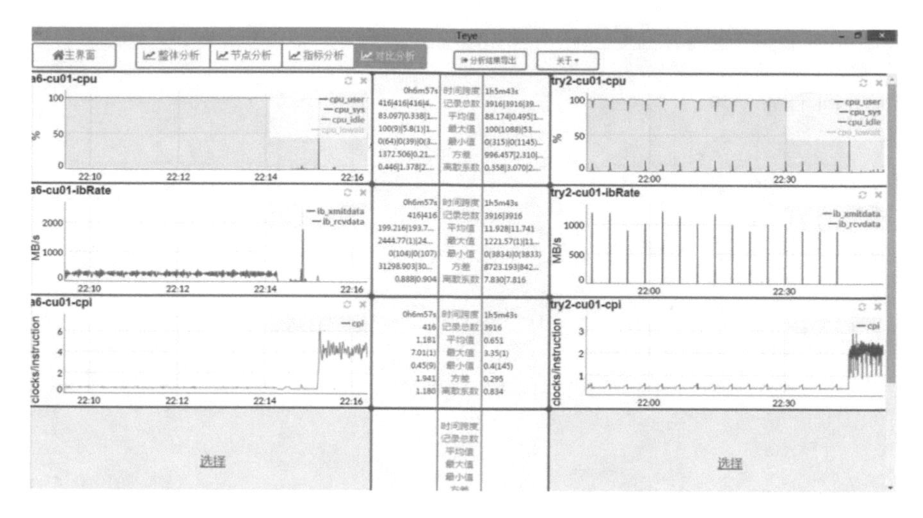

Fig. 5.19 The comparison analysis interface of Teye analysis port

With these powerful functions, Teye has become an indispensable software in the fields of high-performance application software development, performance optimization and performance evaluation. Teye also provides considerable help to high-performance application software developers, application software tuning staff, cluster system performance evaluation personnel and institutes.

Part II
Competition

Chapter 6
An Introduction to International Student Supercomputing Competitions

There are three international supercomputing competitions for undergraduate students including the ASC Student Supercomputer Challenge, the ISC Student Cluster Competition, and the SC Student Cluster Competition.

6.1 The ASC Student Supercomputer Challenge

6.1.1 History

The ASC Student Supercomputer Challenge was initiated by Asia and is now the world's largest student supercomputing competition. China proposed this competition, and organizes it with the supercomputing experts and institutions in Japan, Russia, South Korea, Singapore, Thailand, China Taiwan, China Hong Kong, and other countries and regions. This competition also gets a positive response and support from supercomputing experts and institutions of the USA, Europe, and other countries and regions. The ASC competition aims to provide a platform that promotes the exchange and training of supercomputing young talent in different nations and regions, which could potentially lead to a better development of HPC hardware/software technologies, as well as a better drive for innovation in corresponding industries.

ASC began in 2012, and its influence has been continuously rising. In 2012, there were 27 universities in the competition, and the scope was limited to mainland China; in 2013 the scope was extended to the entire Asia, with a total of 43 colleges and universities participating; and ASC14 in 2014 attracted active participation from up to 82 colleges and universities from around the world, and the participating countries and regions cover Asia, Europe, North America, South America and Africa

ASC Community, *The Student Supercomputer Challenge Guide*,
https://doi.org/10.1007/978-981-10-3731-3_6

6.1.2 Participating Teams

The ASC 2014 competition was held in Sun Yat-sen University. The final competition required each team to design and build a supercomputer within the 3000 W power budget, and use the designed supercomputer to complete the tasks of optimizing the various applications. Moreover, all teams had to compete on the parallel optimization of one given application on Tianhe-2, which is the world's fastest supercomputer located at the Guangzhou Supercomputing Center on the campus of Sun Yat-sen University. As a final step, each team had to give a presentation to complete the competition. Table 6.1 lists the 16 final teams.

Table 6.1 The 16 final teams of ASC14

University	Competition record
Nanyang Technological University, Singapore	The first time participating in an international supercomputing competition
Purdue University, USA	A regular participant of the competition at SC
University of Sao Paulo, Brazil	The first time participating in an international supercomputing competition
Ural Federal University, Russia	The first time participating in an international supercomputing competition
University of Miskolc, Hungary	The first time participating in an international supercomputing competition
Ulsan University of Science and Technology, Korea	South Korean supercomputing competition champion, participated in the final of ASC13
Hong Kong Polytechnic University	Participated in the preliminary stage of ASC13
National Tsing Hua University, China Taiwan	SC10, SC11 champion, ASC13 runner-up
Taiyuan University of Technology, China	The first time participating in an international supercomputing competition
Tsinghua University, China	Host and champion of ASC12 final stage, ISC12 champion, ASC13 champion, and ASC13's highest computing performance award
Zhejiang University, China	The first time participating in an international supercomputing competition
Shanghai Jiao Tong University, China	Host of ASC13 final stage, and ASC13's LINPACK performance runner-up
Beijing University of Aeronautics and Astronautics, China	The first time participating in an international supercomputing competition
Huazhong University of Science and Technology, China	ASC13's MIC application optimization award and ISC13's highest computing performance award
National University of Defense Technology, China	ASC12 runner-up, ISC12, SC12 and SC13 highest performance computing award
Sun Yat-sen University, China	ASC 13 GROMACS application best score

In the end, Shanghai Jiao Tong University (shown in Fig. 6.1) won the final championship, and Nanyang Technological University in Singapore (shown in Fig. 6.2) became the runner-up. Sun Yat-sen University (shown in Fig. 6.3) won the highest computing performance award, and the ePrize computing challenge award that was started at this year's competition was won by Shanghai Jiao Tong University. Shanghai Jiao Tong University, Tsinghua University (shown in Fig. 6.4), Taiyuan University of Technology (shown in Fig. 6.5), Beijing University of Aeronautics and Astronautics (shown in Fig. 6.6) shared the application innovation award.

ASC13 was held in Shanghai Jiao Tong University, and a total of 43 colleges and universities participated in the competition. After a quite competitive

Fig. 6.1 ASC14 champion, ePrize computing challenge award and application innovation award: Shanghai Jiao Tong University

Fig. 6.2 ASC14 runner-up: Nanyang Technological University

Fig. 6.3 The highest computing performance award: Sun Yat-sen University

Fig. 6.4 The application innovation award: Tsinghua University

Fig. 6.5 The application innovation award: Taiyuan University of Technology

Fig. 6.6 The application innovation award: Beijing University of Aeronautics and Astronautics

preliminary stage, Tsinghua University, National University of Defense Technology, Shanghai Jiao Tong University, Huazhong University of Science and Technology, Sun Yat-sen University, Taiwan Tsing Hua University, Chinese University of Hong Kong, University of St. Petersburg (Russia), Ulsan University of Science and Technology (South Korea), King Abdulaziz University (Saudi Arabia) and ten other universities entered the final stage.

In the final stage, ten teams independently built the platform in the field within the power budget of 3 KW, and then competed on five different benchmarks or applications, including LINPACK, GROMACS, BSDE, WRF, and OpenCFD. The total score of the final competition was 100 points, with 18 points for each benchmark, and 10 points for the presentation.

Eventually, Tsinghua University (shown in Figs. 6.7 and 6.8) won both the championship and the highest computing performance award with excellent performance; Taiwan Tsing Hua University (shown in Figs. 6.7 and 6.9) became the runner-up with a fine margin; Huazhong University of Science and Technology (shown in Fig. 6.10) won the MIC application optimization award.

ASC12 was held at Tsinghua University. Among the 27 universities competitors entered: National University of Defense Technology, University of Science and Technology of China, Tsinghua University, Wuhan University, Shandong University and Sun Yat-sen University went into the final competition.

Fig. 6.7 Tsinghua University and Taiwan Tsing Hua University in ASC13

Fig. 6.8 Champion and the highest computing performance award: Tsinghua University

Fig. 6.9 Runner-up: Taiwan Tsing Hua University

Fig. 6.10 MIC application optimization award: Huazhong University of Science and Technology

Fig. 6.11 Champion: Tsinghua University

Fig. 6.12 Runner-up: National University of Defense Technology

Within the limit of 3 kW power, each team had to design and build their own supercomputer system according to their understanding of the supercomputer and the application software. The organizing committee announced the test cases on site, and each team had to optimize five different benchmarks or applications, including HPL, OPENFOAM, CPMD, CP2K and NEMO.

Eventually, Tsinghua University (shown in Fig. 6.11) and National University of Defense Technology (shown in Fig. 6.12) won the first and the second prize of ASC12, respectively.

6.2 ISC Student Cluster Challenge

6.2.1 History

In 1986, Professor Hans Werner Meuer (Director of Computer Center and Professor of Computer Science Department, University of Mannheim, Germany) organized and established the Mannheim Supercomputer Seminar. At that time 81 people participated in the seminar. Afterwards, the seminar has been held every year and has gradually become an international supercomputer conference and exhibition (ISC), which attracts supercomputing experts, vendors and researchers from around the world. Since 1993, this annual conference has become one of the occasions on which the update of the Top500 list is announced. Another conference that announces the new Top500 list is SC (Supercomputing Conference, USA.). Until June 2014 the ISC has been held 29 times and has been continuously run by the Meuer family.

The ISC is a recognized international conference of the Institute of Electrical and Electronics Engincers (IEEE). Generally, ISC provides a technical agenda of five days, and hundreds of experts and scholars from leading technology centers and companies attend the conference. The technical agenda includes: training, seminars, industry links, and sponsor links. ISC 2015 was held in Frankfurt, Germany (Table 6.2).

Since 2009, the ISC organizing committee has expanded its scope and added ISC Cloud and ISC Big data, which were well accepted.

Table 6.2 The year and location of recent ISC conferences

Year	Location	Conference
2015	Frankfurt, Germany	International supercomputing conference
2014	Leipzig, Germany	
2013		
2012	Hamburg, Germany	
2011		
2010		
2009		
2008	Dresden, Germany	
2007		
2006		
2005	Heidelberg, Germany	
2004		
2003		
2002		
1993	Mannheim, Germany	Mannheim supercomputer seminar

6.2.2 Participating Teams

ISC Student Cluster Challenge is a young competition, which has only been held for three years. The first competition was held during ISC12 in Hamburg, Germany. Tsinghua University and National University of Defense Technology (shown in Fig. 6.13) from China, the University of Colorado and New York State University at Stony Brook from the USA, and the Karlsruhe Institute of Technology of Germany attended the competition.

The competition included one benchmark (LINPACK), and six applications (OPENFOAM, CPMD, CP2K, NEMO, WRF and GROMACS). The performance of the benchmark and applications on the built system accounted for 80% of the total score.

In the end, the team from Tsinghua University from China won the championship, which is the first international supercomputing champion won by students from mainland China. Meanwhile, the team from National University of Defense Technology won the highest computing performance award. The New York State University at Stony Brook team won the debug award, due to their excellent performance on debugging their equipment. The team from Karlsruhe Institute of

Fig. 6.13 **a**, **b** National University of Defense Technology and Tsinghua University won awards; **c**, **d** National University of Defense Technology and Tsinghua University built systems on the competition site

Technology was elected the most popular team, with the highest number of votes from Facebook and Twitter.

ISC13 was held in the Leipzig Conference Center on June 16–19, 2013 (shown in Fig. 6.14). Eight teams from China, the USA, Germany, the United Kingdom, South Africa and Costa Rica participated in this supercomputing competition. The eight teams were the Karlsruhe Institute of Technology, Chemnitz University of Technology, Tsinghua University, Huazhong University of Science and Technology, University of Colorado, Purdue University, The University of Edinburgh and the South African Centre for HPC.

After an intensive competition of four days, the team from the South African Centre for HPC won the championship, with Tsinghua University as the runner-up.

Fig. 6.14 **a** The ISC13 award ceremony; **b**, **c** the teams of Tsinghua University and Huazhong University of Science and Technology

Huazhong University of Science and Technology from China won the highest LINPACK performance award (see Fig. 6.14). In this competition, the interview counted for 40% of the total score. Although the Tsinghua team was ahead of the South African team by over ten points for the performance score, the South African team won the championship with a better performance at the interview. Therefore, improving the presentation skills of the Chinese teams in international competitions is another aspect that requires attention.

Fig. 6.15 **a** The ISC14 award ceremony; **b**, **c** Shanghai Jiao Tong University and Tsinghua University, at the competition site

The ISC14 Student Cluster Challenge was held in Leipzig, Germany (shown in Fig. 6.15a). Participating teams were:

- China: Shanghai Jiao Tong University (shown in Fig. 6.15b), Tsinghua University (shown in Fig. 6.15c), and the University of Science and Technology of China.
- South Africa: The Centre for HPC.
- The USA: University of Colorado, and the alliance of Massachusetts Institute of Technology, Bentley University and Northeastern University.
- South Korea: Ulsan University of Science and Technology.
- The United Kingdom: The University of Edinburgh.
- Germany: Chemnitz University of Technology and the University of Hamburg.
- Brazil: the University of São Paulo.

Among these teams, Shanghai Jiao Tong University, Tsinghua University, University of São Paulo, Brazil, and South Korean Ulsan University of Science and Technology were from the ASC Student Supercomputer Challenge.

This competition included the optimization of the conventional LINPACK benchmark, and openFOAM (computational fluid dynamics software package), GADGET (celestial cosmological N-system simulation software) and Quantum ESPRESSO (nano-scale electronic structure and material modeling software). Moreover, two mystery tasks were announced on site: HPCG (high-performance conjugate gradient benchmark) optimization, and the system efficiency and stability test via Quantum ESPRESSO software to be completed within 20 min. For the Quantum Espresso test, apart from the requirements of accuracy, the team that accomplished the test with the lowest power consumption got full points. The other teams that finished the test would be given points according to the ratio of their power consumption to the lowest power consumption that the winning team achieved. This task emphasized the power optimization within the given job and time requirements, which demonstrated the community's focus on performance power ratio when running practical applications. It was the first time that this kind of mystery task and rules were introduced to the competition, which challenged each team's capabilities on building supercomputer systems, balancing performance and power consumption, cooperating and on-site adaptabilities. This task attracted the most attention during the ISC14 competition.

After intensive preparation and competition, the South African team from the Centre for HPC got the full 16 points with the lowest 1700 W peak power consumption. Teams from Chinese Tsinghua University and Shanghai Jiao Tong University got 11.33 points and 10.07 points respectively with the peak power consumption of 2400 and 2700 W. They were the only three teams that successfully completed the task. Unfortunately, due to various reasons, the other eight teams failed to complete this mystery task.

Eventually, the South African team from Centre for HPC won the championship with an excellent and balanced performance. The University of Science and Technology of China and Tsinghua University won the second and third prizes.

After the competition, the Tsinghua captain Kaiwei Li made a comment that enjoying the competition was more important than winning the prize. This showed that inspiring young people's interest in supercomputing was undoubtedly a very important part of the competition.

Table 6.3 lists the judges of ISC13 and ISC14. The list is stable with a majority of them from the HPC Advisory Council (HPCAC) and ISC organizing committee. We have not yet seen judges from Chinese institutions.

6.2.3 Patterns and Trends

The ISC Student Cluster Challenge aims to introduce the next generation of students to the HPC (supercomputing) field. The chairman of HPC Advisory Council (the organizer of the competition), Gilad Shainer, introduced that more than 20 applications had been received for ISC13. He hoped that this competition would become an attractive learning opportunity and also a fascinating opportunity for young talented students to demonstrate their knowledge and skills. The HPC Advisory Council receives stable funding support from Mellanox, which assures the continuity of this competition.

Table 6.3 The list of judges of ISC13 and ISC14

ISC13	ISC14
Gilad Shainer (HPC Advisory Council)	Gilad Shainer (HPC Advisory Council)
Brian Sparks (HPC Advisory Council)	Brian Sparks (HPC Advisory Council)
Hans Meuer (ISC)	Thomas Meuer (ISC)
Martin Meuer (ISC)	Martin Meuer (ISC)
Gerd Buettner (Airbus)	Gerd Buettner (Airbus)
Tong Liu (HPC Advisory Council)	Tong Liu (HPC Advisory Council)
Pak Lui (HPC Advisory Council)	Pak Lui (HPC Advisory Council)
Marcel Schneider (ISC)	Marcel Schneider (ISC)
Shelley Smith (ISC)	Shelley Smith (ISC)
Dan Olds (Gabriel Consulting Group)	Dan Olds (Gabriel Consulting Group)
Happy Sithole (Center for High Performance Computing, South Africa)	Happy Sithole (Center for High Performance Computing, South Africa)
Doug Smith (University of Colorado)	Doug Smith (University of Colorado)
Hussein Harake (HPC Advisory Council Switzerland Center of Excellence)	Hussein Harake (HPC Advisory Council Switzerland Center of Excellence)
Sadaf Alam (Swiss Supercomputing Center)	Sadaf Alam (Swiss Supercomputing Center)
Cydney Stevens (HPC Advisory Council)	Martin Hilgeman (HPC Advisory Council)
Christophe Mion (Scalable Graphics)	Christophe Mion (Scalable Graphics)
Filippo Spiga (University of Cambridge)	Volker Springel (Heidelberg University)

Looking back to the competition in the last four years, the organizing committee, the judges and the competing teams have all made progress. The ISC also regularly follows the ASC and SC competitions. For instance, ISC14 introduced the same awards as the ASC14 awarded.

Due to the influence and the coverage of the ISC conference, the ISC competition has its inherent international influence. However, if the committee cannot balance the fairness and the coverage well, the long-term development of the competition may have various issues.

6.2.4 Further Reading

- **ISC** See more information via http://www.isc-events.com/
- **ISC12** Find more information via http://www.isc-events.com/isc12/student-programs.html
- **ISC13** See more information via http://www.isc-events.com/isc13/student-cluster-challenge.html
- **ISC14** Find more information via http://www.isc-events.com/isc14/student-cluster-competition.html

6.3 SC Student Cluster Competition

6.3.1 History

The SC supercomputing conference is the annual meeting of US supercomputing community. Along with the increasing number of participants, and the expanding scope, SC has become the most famous annual event of HPC. The first session of the SC conference was held in Orlando, USA, in November 1988. The number of participants was just over 100. By 2013, the number of participants has reached nearly 10,000. SC is jointly funded and held by ACM Sigarch, ACM Sighpc and the IEEE Computer Society.

The SC annual conference is rich in content, and usually includes invited keynote speakers, technical sessions for papers and posters, training, exhibitions, seminars, awards and new technology sharing. Each activity has a detailed plan to aid management. The annually increasing number of participants and influence show the high recognition of the quality of the SC conference.

During SC05, several experts contemplated whether students could utilize simple tools to build a supercomputer. Based on such an idea, the first SC student cluster competition was started at SC07 in Reno. Since then, this competition continued for seven years. SC is the birthplace of student supercomputing

	SC13 Denver, CO November 18-21, 2013		SC12 Salt Lake City, UT November 10-16, 2012
	SC11 Seattle, WA November 12-18, 2011		SC10 New Orleans, LA November 13-19, 2010
	SC09 Portland, OR November 14-20, 2009		SC08 Austin, TX November 2008
	SC07 Reno, NV November 10-16, 2007		SC06 Tampa, FL November 2006

Fig. 6.16 The locations of the 2006–2013 SC conferences

competition, and has led to the ISC, ASC and other regional competitions in South Korea, Taiwan and South Africa. Figure 6.16 shows the locations of the 2006–2013 SC conferences.

6.3.2 Participating Teams

Due to the space limitations, we only discuss two SC competitions here.

The SC12 competition was held in October of that year in Salt Lake City. The participating teams included: Boston University, Taiwan Tsing Hua University, the University of the Pacific, the University of Texas, National University of Defense Technology, the University of Science and Technology of China, Texas Tech University and Purdue University.

The SC12 competition conducted a total of seven major tasks and 24 minor tasks. In addition to the common LINPACK test, it also included applications, such as LAMMPS, QMCPACK, CAM and PFLOTRAN.

The University of Texas team won the championship with a score of 79.86. The team of National University of Defense Technology won the LINPACK award, as shown in Fig. 6.17.

The SC13 competition was held in Denver, Colorado, US. Sponsored by the Inspur, National University of Defense Technology team participated in the race on behalf of China. Other teams came from the University of Tennessee at Knoxville, the University of Colorado Boulder Campus, the University of the Pacific, the University of Texas at Austin, Massachusetts Green High Performance Computing

Fig. 6.17 The team of National University of Defense Technology won the LINPACK award

Center and Friedrich-Alexander University of Erlangen-Nuremberg. There was also a joint team from five Australian universities.

The competition included tuning tasks for WRF, NEMO5, GraphLab and a mystery application. GraphLab is a very popular open-source project, and GraphLab developers constantly pursue the innovation and development of graph computing techniques in order to enable massive data processing. The SFrame's appearance seems low-key and mysterious, but its functions should not be overlooked. It has extended GraphLab to the processing of data forms and tables, and can easily handle TB-scale data.

In the end, National University of Defense Technology team won the SC13 highest computing performance award with 8.22 TFlops of LINPACK performance. This is also the third time that National University of Defense Technology has won this award after ISC12 and SC12. The organizing committee also

Fig. 6.18 The team from the University of Texas won the championship

particularly noted at the awards ceremony that the team was from the same institute that designed and built Tianhe-2, the top system on the then Top500 list. The team from the University of Texas at Austin (shown in Fig. 6.18) won the SC13 championship.

6.3.3 Patterns and Trends

As the most influential supercomputing conference, SC has an unrivaled influence and international recognition in the supercomputing community. Dustin Leverman, the SC13 and SC14 Student Competition President and Oak Ridge National Laboratory system director, was a former participant that benefitted from the SC07 competition, and hoped to bring such a good opportunity to more new students. Brent Gorda, the general manager of the Intel high-performance data department, happily expressed that participating in this competition could promote the academic-industry cooperation, and bring more job opportunities for students.

With the influence of the SC student cluster competitions, many students got internships from the well-known HPC companies, and gained knowledge about high-performance computing. The competition also encouraged students that had participated in the competition to actively communicate with other students that have no chance of participation. Many colleges and universities subsequently set up HPC-related courses, such as the University of Colorado at Boulder, and the Sun Yat-sen University.

6.3.4 Further Reading

- **SC** See more information via http://supercomputing.org/
- **SC12** See more information via http://sc12.supercomputing.org/content/student-cluster-competition.html
- **SC13** See more information via http://sc13.supercomputing.org/content/student-cluster-competition

Chapter 7
History and Prospects of ASC Student Supercomputer Challenge

7.1 The First Chinese Student Supercomputer Challenge

7.1.1 The First International Competition in China

In 2012, with the guidance of the Ministry of Science and Technology of China and the Ministry of Education of China, and with the joint support of the Inspur Group, the International Supercomputing Conference Organizing Committee (ISC), and the international HPC Advisory Committee (HPC-AC), the "Student Supercomputer Challenge and ISC Student Cluster Challenge Regional Preliminary Stage" was introduced into China for the first time. This competition received strong support from the Chinese government and the scientific community in China. Yilian Jin, Xubang Shen, Xin-gui He and Zuoning Chen, four senior academicians of the Chinese Academy of Engineering, served as the chairman and vice-chairmen of the competition steering committee. This competition also received full support from the Department of High and New Technology Development and Industrialization of Ministry of Science and Technology of China, and Tsinghua University.

7.1.2 The Competition Process

Since the announcement of the event in February 2012, teams from 27 universities registered before the deadline of March 2. After a careful evaluation of the proposals from the 27 teams, the National University of Defense Technology, the University of Science and Technology of China, Tsinghua University, Wuhan University, Shandong University and Sun Yat-sen University were selected for the finals. On April 16, these six teams gathered in Beijing for the final competition. On April 19, according to the performance results and the presentation scores given by

ASC Community, *The Student Supercomputer Challenge Guide*,
https://doi.org/10.1007/978-981-10-3731-3_7

the competition committee, Tsinghua University and National University of Defense Technology eventually won the entrance tickets for the ISC12 Student Cluster Challenge held in June in Hamburg, Germany.

7.1.3 The Competition's Three Key Elements: College Students, International Standards and Practical Skills

The goal of the competition was to examine college students' practical supercomputing skills under international rules and benchmarks. Each university was only allowed one participating team, with six students and one guiding faculty. Each participating student must be a full-time undergraduate student in that university.

Within the power budget of 3 kW, each team had to build their own supercomputer system according to their understanding of a supercomputer and the application software. The organizing committee announces the benchmark and application tasks at the time and each team has to test and optimize the given applications, and finally to give a presentation on site. This competition is practicality oriented: using practical benchmarks and applications, such as HPL, OPENFOAM, CPMD, CP2K and NEMO.

7.2 Asian Student Supercomputer Challenge 2013 (ASC13)

7.2.1 ASC13's Orientation and Purpose

The foundation of ASC Student Supercomputer Challenge was proposed by China, and organized by supercomputing experts and institutions in Mainland China, Japan, Russia, South Korea, Singapore, Thailand, China Taiwan, China Hong Kong and other countries and regions. The Inspur Group sponsored the competition. The ASC competition aims to provide a platform that promotes of the exchange and training of supercomputing young talent among different nations and regions, which could potentially lead to a better development of HPC hardware/software technologies, and a better drive for innovation in corresponding industries.

As the first international supercomputing competition in Asia, ASC13 has attracted attention and positive responses from many Asian universities. The total 43 participating universities covered East Asia, West Asia, North Asia and South Asia, including China, Russia, South Korea, India, Kazakhstan, Saudi Arabia, China Taiwan, China Hong Kong and many other countries and regions.

The participators included experienced teams such as the team from National Defense Science and Technology University, which had won the highest computing performance award in SC12 Student Cluster Competition, teams from South

Korean Pukyong National University, Moscow Bauman State Technical University, Indian University of Mumbai, India, Kazakhstan National University, King Abdulaziz University and other famous universities from Asian countries or regions, and teams from Nanjing University, Wuhan University, Tongji University, Fudan University, Northwestern Polytechnical University and other well-known universities from Mainland China.

7.2.2 The Competition Rules, Schedules and Organizational Structures

The ASC13 competition was organized by the Inspur Group, with support from the ISC organizing committee and international HPC Advisory Committee, and jointly co-hosted by the State Key Laboratory of High-end Sever & Storage Technology and Shanghai Jiao Tong University. This competition also received strong support from the Department of High and New Technology Development and Industrialization of Ministry of Science and Technology of China, Department of Science and Technology of Ministry of Education of China, Shanghai Supercomputer Center, Computer Network Information Center of Chinese Academy of Science and the National Supercomputer Center in Tianjin, Jinan, Changsha and Shenzhen.

The competition was divided into three stages. The first stage was the university enrollment. The universities organized the participating teams, and the registration ended on January 25. The second stage was the preliminary round, and it took place in February. According to competition requirements, all teams had to submit a written proposal, and the review committee then selected the finalists. The third stage was the final competition that was held in April. With the supercomputer equipment provided by the Inspur the final teams had to build their own supercomputer system according to their understanding of the supercomputer and the application software. Using the supercomputer they had constructed the teams then tested and optimized the given applications, and presented the competition results on site. The judges then ranked the teams according to the performance and presentation results. According to the rules, each university can only have one team, and the six students of the team must all be undergraduates.

In addition to the performance tuning tasks on the built systems, this competition also included a MIC optimization task for the BSDE program used in option pricing, which showed the Asian feature of this competition.

7.2.3 The Ten Finalists of ASC13

The ten teams that entered the final stage of ASC13 all had rich experience in high-performance computing:

King Abdulaziz University (KAU; Saudi Arabia): The development of supercomputers in the Kingdom of Saudi Arabia has attracted a large amount of attention from the King and the government. KAU have a strong research force for the supercomputing applications in oil exploration, life science, climate modeling, and so on.

Saint-Petersburg State University (Russia): Saint-Petersburg State University has started a number of HPC-related courses. The participating students representing the Saint-Petersburg University all have received training in high-performance computing.

Ulsan National Institute of Science and Technology (South Korea): Ulsan National Institute of Science and Technology values supercomputing education. The university organizes parallel computing courses both in winter and summer, and attracts many students to participate. The participating students had already achieved outstanding results in the KSC student supercomputer challenge before the ASC13 competition.

Taiwan Tsing Hua University (China Taiwan): National Tsing Hua University has won the SC Student Cluster Competition championship quite a few times, and has attracted the attention ofthe supercomputing community home and abroad.

Chinese University of Hong Kong (China Hong Kong): Chinese University of Hong Kong has built a solid supercomputing education model. The university has set up a computer science center, serving the university's interdisciplinary research related to supercomputing. Also, the university has supercomputing courses for undergraduates and graduates.

Tsinghua University: Tsinghua University owns a 172 TFlops supercomputer, which was jointly developed with the Inspur Group. In 2012, the Tsinghua University team won the ISC12 Student Cluster Challenge championship, and this outstanding achievement surprised the supercomputing community.

Shanghai Jiao Tong University: Shanghai Jiao Tong University also has a solid supercomputing education model. The university provides undergraduate supercomputing courses and has invited internationally well-known high-performance computing scholars to teach on its courses. This university has also set up a high-performance computing scholarship to reward outstanding students. ASC13 competition was hosted at the Center for High Performance Computing of Shanghai Jiao Tong University, and received strong support from SJTU Network & Information Center and School of Electronic Information and Electrical Engineering. With the students' enthusiasm, not only a six-student team was quickly formed, but also it attracted more and more students to participate in the learning and application of supercomputing technologies.

National University of Defense Technology (NUDT): They have won both the ISC12 and SC12 highest computing performance awards, and continuously made new records for this award. NUDT built the Tianhe-1A and Tianhe-2 systems, both of which are top supercomputers in the world.

Huazhong University of Science and Technology: The Center for High Performance Computing of Huazhong University of Science and Technology has built many high-performance-computing-related courses, such as parallel processing and parallel programming, and has regularly organized training events for the entire university. Lab students have spontaneously formed a high-performance computing interest group, and have participated in some supercomputing-related competitions (such as the financial computing competition), and achieved good results.

Sun Yat-sen University: Sun Yat-sen University owns the leading supercomputing platform in the southern Chinese universities. This university has paid considerable attention to applying supercomputing technologies to the development of new disciplines and research innovation. The university has always focused on training. Related courses, training events, special interest groups and other forms of study are integrated to promote the HPC community within the university.

7.2.4 The Jury Committee's Comprehensive Evaluation on the Participating Universities

The jury committee spoke highly of the teams participating in ASC13.

The rigorous and systematic methodology: In terms of the overall design submitted by different universities, these teams had an in-depth understanding of high-performance computing. From the optimization of the system platform to the optimization of the actual application software, to the code acceleration on the many-core platforms, these teams all demonstrated a rigorous and systematic methodology and superb technical skills. Also, the conclusions demonstrated at the presentations were a good summary of their efforts.

A mature system design: The system designs proposed by a number of teams were quite mature, and included innovative architectures. They took full account of the characteristic differences, and the performance/power ratios, of different applications. These in-depth analyses helped to maximize the potential performance of the machine.

Excellent performance testing and tuning: The testing and tuning strategies submitted by different universities all provided a detailed description of the system, hardware and software configurations. The proposed tuning method took a comprehensive approach to explore different factors of the system, and achieved significant performance improvement. For the option pricing application BSDE all the teams provided a detailed analysis and hot-spot test of the program, and designed highly efficient MIC programs with a high speed-up.

Deep optimizations of the algorithm and the code: Many teams had a good understanding of both the software and hardware architectures of the cluster, as well as the system and compiler optimization. Some proposed solutions and considered potential problems and the corresponding solutions. A number of teams managed to perform a detailed analysis of the algorithm and code, and to make algorithm-level and code-level optimizations.

7.3 Student Supercomputer Challenge 2014 (ASC14)

7.3.1 Eighty-Two Participating Universities, the Largest Scale in History

The 2014 Student Supercomputer Challenge (ASC14) was jointly organized by the Asian Super Computer Society, Sun Yat-sen University and the Inspur Group. Competition registration started in November 2013, and the preliminary competition was held in January–February 2014. The final competition was held in Sun Yat-sen University, Guangzhou, from April 21 to April 25, 2014.

The 82 participating teams were from North America, South America, Africa, Asia and Europe, setting a new record for the largest number of participants and the widest coverage of countries and regions in global supercomputing competitions.

Among the 82 participating universities, there were some teams with excellent experience in supercomputing competitions. Tsinghua University had consecutively won the ASC12, ISC12 and ASC13 championships. Sun Yat-sen University had a strong capability in application optimization and achieved the best optimization results of GROMACS in ASC13. National University of Defense Technology won SC and ISC's highest computing performance award three times. Huazhong University of Science and Technology obtained ASC13 MIC application optimization award and ISC13 highest computing performance award. Taiwan Tsing Hua University has won the SC student cluster competition championship twice, and was the runner-up of ASC13. Purdue University and the University of Colorado were both regular participants of the SC student cluster competition. South African University of the Western Cape, and South Korean Ulsan University of Science and technology were respectively champions of South Africa and South Korea national supercomputer competitions.

7.3.2 Gathering of International Supercomputing Experts

The committee of experts of ASC14 was jointly chaired by Prof. Jack Dongarra, who was the major contributor of LINPACK and initiated the TOP500 list, and Dr. Dona L. Crawford, who is the deputy director of Lawrence Livermore National Laboratory. Academician of Chinese Academy of Sciences Xubang Shen, Academicians of the Chinese Academy of Engineering Xin-gui He and Chen Zuoning, ACM Fellow, Tokyo Institute of Technology Professor Satoshi Matsuoka, and Russian Academy of Sciences Fellow, Computational Science Director Abramov, Sergei M all served as vice-chairmen.

Committee members included Depei Qian (professor of Beijing University of Aeronautics and Astronautics, director of Sino-German Joint Software Institute, leader of 863 Program committee of experts), Xiangke Liao (professor, dean of College of Computer, National University of Defense Technology), Hai Jin (professor, dean of School of Computer Science & Technology, Huazhong University of Science and Technology, leader of ChinaGrid experts group), Guangming Liu (director of National Supercomputer Center in Tianjin), Jiachang Sun (professor of Institute of Software in Chinese Academy of Sciences), Guoxing Yuan (professor of Beijing Institute of Applied Physics and Computational Mathematics), Zili Xi (director of Shanghai Supercomputer Center), Weidong Gu (director of National Supercomputer Center in Jinan), Yueguang Zhang (deputy director of National Supercomputer Center in Shenzhen), Guangwen Yang (professor, director of the HPC institute, Tsinghua University), Yunquan Zhang (professor of Institute of Software in Chinese Academy of Sciences), Kenli Li (director of National Supercomputer Center in Changsha), Eric YH Tsui (vice president of Hong Kong Knowledge Management Society, professor of Hong Kong Polytechnic University), Taisuke Boku (professor of Computer Science Department, University of Tsukuba, Japan), Lee Yoshio Oyanagi (dean of Information School, Nippon Institute of Technology, Japan), Yoshio Oyanagi (chairman of the Information Department, Kogakuin University, Japan), Joon KISTI Woo (South Korea), Michalewicz Marek (senior director of Singapore A*Star Computing Resource Center), Simon See (consultant of Shanghai Jiao Tong University Supercomputer Center), Putchong Uthayopas (chairman of Computation Science and Engineering Center, Kasetsart University, Thailand), Fortes (director of ACIS Laboratory), Earl C. Joseph II (vice president of IDC), Rajeev Wankar (professor of Hyderabad University, India), José AB Fortes (US AT & T distinguished scholar), Walter Binder (deputy director of STFC Laboratory of Computer Science and Engineering Center, associate professor of School of Information, University of Lugano, Switzerland), Peter Arzberger (executive director of the American NPACI Center) and Kwan Wing Keung (assistant governor of Hong Kong University IT Center).

The steering committee was chaired by Yuhai Zhao (chief of Department of High and New Technology Development and Industrialization, Ministry of Science and Technology of China). The committee members included Xianwu Yang (vice chief of Department of High and New Technology Development and Industrialization, Ministry of Science and Technology of China), Endong Wang (one of the initiators of ASC, director of State Key Laboratory of High-end Server & Storage Technology, senior vice president of the Inspur Group), Weimin Zheng (professor of Tsinghua University, chairman of China Computer Federation), Xuebin Chi (director of Supercomputing Center, Computer Network Information Center, Chinese Academy of Sciences), ChingPak-Chung (professor of Chinese University of Hong Kong, vice dean of Shun Hing Advanced Engineering Research Institute), Martin Meuer (CEO of the International Supercomputing Conference) and Gilad Shainer (the chairman of HPCAC advisory committee).

Zeyao Mo, the deputy director and researcher of Beijing Institute of Applied Physics and Computational Mathematics, chaired the ASC14 jury committee.

7.3.3 Preliminary Competition: Four Tasks for Selecting the Finalists

In the ASC14 preliminary competition, the committee specified four tasks. Each team had to design its own supercomputing system within the power budget of 3000 W, and complete the testing and optimization of three different benchmarks or applications, including HPL, Quantum Espresso and 3D-EW.

Firstly, the teams needed to demonstrate their capabilities to design and build a supercomputing system. The participating teams had to design the supercomputing system based on their considerations of the computing performance, power control, high-speed network, accelerator and software environment. The validity and innovation were the most important metrics that the jury committee used to assess the proposed system.

The HPL test, which is the most important metric of the TOP500 supercomputer ranking, mainly examined the team's ability to optimize the floating-point performance of the hardware platforms. To achieve good results the participating teams had to conduct comprehensive research on the supercomputer system architecture, multi-level memory consistency, high-speed network and the numerical algorithm, explore different options of homogeneous systems and heterogeneous systems, and perform iterative optimizations based on performance and power consumption results. The jury committee specifically encouraged teams to optimize the HPL software at the algorithm and code levels.

The Quantum Espresso software is a widely used open-source quantum molecular dynamics modeling software. Quantum Espresso contains the PWscf, FPMD, CP, PWgui and Atomic modules, as well as the electronic self-consistent calculation, lattice dynamics calculation, subsequent data processing, electronic transport properties calculation and molecular dynamics modules. The Quantum Espresso software can simulate the performance of surreal materials, and is mainly used in the development of new materials. Currently, other quantum molecular dynamics software usually has a weakness of poor scalability, while Quantum Espresso can scale to thousands of or even tens of thousands of cores according to different modeling scenarios.

3D-EW (a simulation method of three-dimensional elastic wave equation with P- and S-wave separation) was the ASC14 MIC application optimization task, and the participators had to complete the application optimization on the CPU+MIC heterogeneous clusters constructed by the organizing committee. 3D-EW is a Chinese self-developed application software selected from the candidates of the "Call for applications of China" event, which started at the same time as ASC14 enrollment. This software is independently developed by BGP INC., China National Petroleum Corporation, and has been widely used in oil and gas exploration.

7.3.4 Final Competition: Seven Tasks Challenging the Finalists' Skill Limit

The 16 finalists had to design and build a supercomputer within a 3000 W power constraint, and use the supercomputer to complete the mission of optimizing the required benchmarks or applications. Moreover, all teams had to compete on the parallel optimization of one given application on Tianhe-2, which is located at the Guangzhou Supercomputing Center on the campus of Sun Yat-sen University. At the end each team gave a presentation about their system and the performance tuning results.

In addition to the four tasks in the preliminary round, the final stage of ASC14 added LICOM and SU2, with a mystery application that was announced on the final day.

LICOM is an ocean circulation model that was independently developed by the Institute of Atmospheric Physics, China Academy of Sciences. It is an Eta coordinate primitive equation ocean circulation model, and is also the ocean component of the LASG atmosphere-ocean coupled model. LICOM presents well the major characteristics of the large-scale ocean circulation. LICOM is the third generation of the global ocean circulation model developed by LASG scientists, and has been widely used to study ocean circulation, climate change modeling, atmosphere-ocean interaction and other related topics.

SU2 is an open-source computational fluid mechanics software, developed by Stanford University. SU2 is mainly used for solving partial differential equations, as well as aerodynamic research and aircraft shape optimization. SU2 supports parallel computing. Moreover, SU2 adopts the C++ language with high readability.

7.3.5 Global Young Talents Challenging the Performance Limit of the World's Fastest Supercomputer

In the ASC14 final competition, each participating team competed on optimizing the 3D-EW application on the world's fastest supercomputer—Tianhe-2—to challenge Tianhe-2's performance limit.

As an application optimization task, 3D-EW has a strong parallelism, and retains a large optimization space for the teams to explore. The participating teams can make a full use of their imagination and innovation and optimize the software performance while scaling to the large-scale heterogeneous clusters.

7.3.6 Dark Horse Emerging

Although it was the first time that Nanyang Technological University (Singapore) and Taiyuan University of Technology had participated they performed extremely well, and were ranked the first in the preliminary stage at home and abroad respectively. They both became the dark horses in this competition.

Nanyang Technological University exhibited a profound understanding of condensed matter physics in the Quantum Espresso optimization, and optimized the code from the perspective of the underlying physics.

Taiyuan University of Technology obtained the best results of the 3D-EW parallel optimization on the CPU+MIC heterogeneous parallel platform. They dramatically reduced the application run time from the original serial 9000 s to the parallel 21 s, and achieved an amazing speedup of 400 times on the four compute nodes.

It was not a fluke for Taiyuan University of Technology to get the best MIC application optimization. According to the MIC login log collected by the organizing committee, Taiyuan University of Technology reached a total of 500 h, which was the highest among the 82 universities. From January 2 (preliminaries started) to February 27 (deadline to submit preliminary results), the preliminary stage lasted 56 days, a total of 1344 h, including the Spring Festival, the Lantern Festival, and the winter break. This meant that Taiyuan University of Technology team spent more than one-third of the total time on the tuning of the application on MIC.

7.3.7 ePrize Computing Challenge Award for Motivating Global Supercomputing Talent

To encourage the development of supercomputing young talent, the ASC committee announced the introduction of the ePrize computing challenge award for the team who got the highest scalability and performance optimization of the application in the competition. This award aimed to encourage young talent to solve the hard problems in supercomputing applications and to challenge the computing performance limits. It also aimed to encourage the young talent to utilize the supercomputer to achieve breakthroughs in the science and engineering applications.

"e" is the most important natural constant in science, and also represents the next expected scale of the supercomputing performance (exascale). ASC hopes the ePrize computing challenge award will become the Gordon Bell Prize for young talent.

7.4 The Asia Student Challenge 2015 (ASC15)

7.4.1 A Historical One with 152 Teams from Six Continents

ASC15 attracted 152 teams from 135 countries from places that included North America, South America, Africa, Asia and Europe. A new history was made for so many teams, countries and states.

Among the countries, six were new to the competition, including Japan, Germany, the Netherlands, India and Indonesia. At the same time, top universities in the world started to be seen, as shown in Table 7.1.

7.4.2 A Gathering of Specialists from Around the World

Listed below are the members of the Expert committee of ASC 15:

Chair

Dongarra, Jack, Fellow of the AAAS, ACM, SIAM and the IEEE, and a member of the National Academy of Engineering. Distinguished Professor at the University of Tennessee.

Dona L. Crawford, Associate Director Computation, Lawrence Livermore National Laboratory.

Vice chair

Shen, Xubang, Fellow of Chinese Academy of Science.

He, Xingui, Fellow of Chinese Academy of Engineering.

Chen, Zuoning, Fellow of Chinese Academy of Engineering.

Matsuoka, Satoshi, ACM Fellow, Prof. of Tokyo Tech, Japan.

Table 7.1 ASC Top Universities

University	Ranking
Massachusetts Institute of Technology	No. 1 of the world
Nanyang Technological University, Singapore	No. 2 of Singapore
Tsinghua University	No. 1 of China
Bandung Institute of Technology	No. 2 of Indonesia
Univeriti Putra Malaysia	No. 5 of Malaysia
Saint-Petersburg State University	No. 2 of Russia
University of Saint Paul	No. 1 of Brazil
Leiden University	No. 2 of the Netherlands
Tokyo Institute of Technology	No. 4 of Japan
National Tsinghua University	No. 2 of Taiwan, China
The Chinese University of Hong Kong	No. 2 of Hong Kong, China

Source Data from QS University Ranking 2015

Abramov, Sergei M., Director of Theoretical Computer Sciences, corresponding member of the Russian Academy of Science.

Members

Qian, Depei, Professor in Beihang University, Director of the Institute of Advanced Computing Technology. General Chief of High Performance Computer and Network Grid Service Environment (863 key project).

Liao, Xiangke, Professor in National University of Defense Technology, Dean of computer science department.

Jin, Hai, Profess in Huazhong University of Science & Technology, Dean of computer science department. The General Chief of ChinaGrid.

Liu, Guangmin, Director of National Supercomputing Center Tianjin.

Sun, Jiachang, Professor in Institute of Software, Chinese Academy of Sciences.

Yuan, Guoxing, Professor in Institute of Applied Physics and Computational Mathematics.

Zhou, Ximing, Director of Shanghai Supercomputing Center.

Yuan, Xuefeng, Director of Guangzhou Supercomputing Center.

Xi, Zili, Professor of Shanghai Supercomputing Center.

Gu, Weidong, Director of National Supercomputing Center Jinan.

Yang, Guangwen, Director of High Performance Center in Tsinghua University.

Zhang, Yunquan, Professor in Institute of Software, Chinese Academy of Sciences.

Li, Kenli, Director of National Supercomputing Center Changsha.

Hu, Leijun, CTO in Inspur Group.

Tsui, Eric Y.H., Vice President of Hong Kong Knowledge Management Society and Professor in Hong Kong Polytechnic University.

Boku, Taisuke, Professor, Department of Computer Science, University of Tsukuba.

Oyanagi, Yoshio, Dean, Faculty of Informatics, Kogakuin University.

Woo, Joon, Department of supercomputing infrastructure operation, National Institute of Supercomputing and Networking, KISTI.

Michalewicz, Marek, Senior Director of Computer Resource Center A*Star.

See, Simon, Professor, Center for HPC, Shanghai Jiao Tong University and Scientific Computing Advisor, BGI.

Uthayopas, Putchong, Professor and Head of Department of Computer Engineering, President of Computational Science and Engineering Association, Kasetsart University.

Earl C. Joseph II, IDC Program Vice President and Executive Director HPC User Forum.

Rajeev Wankar, Associate Professor in School of Computer and Information Sciences, University of Hyderabad, & Chairman, Placement Guidance and Advisory Bureau, UoH P.O. Central University Hyderabad, India.

José A.B. Fortes, Professor and AT&T Eminent Scholar, Director, Advanced Computing and Information Systems (ACIS) Laboratory, Director, NSF Cooperative Industry-University Cloud and Autonomic Computing Center (CAC).

Mike Ashworth, Associate Director of the Computational Science & Engineering Department at STFC's Daresbury Laboratory.

Walter Binder, associate professor at the Faculty of Informatics of the University of Lugano (USI).

Peter Arzberger, Executive Director, National Partnership for Advanced Computational Infrastructure (NPACI).

Kwan Wing Keung, Asst. IT Director, Information Technology Services, the University of Hong Kong (HKU).

Filippo Spiga, HPC Applications Specialist–University of Cambridge, Interim Executive Director–Quantum ESPRESSO Foundation.

Leverman, Dustin, HPC Storage Administrator, Oak Ridge National Laboratory, Chair of the student supercomputer challenge on SC14.

Umesh Gupta, Research Fellow, Indian Institute of Technology (IIT) Bombay, India.

The Chair of the Evaluation committee
Mo, Zeyao, Vice Director and Professor in Beijing Institute of Applied Physics and Computational Mathematics.

7.4.3 ASC Preliminary: Challenge Covering Molecule to the Universe

ASC15 involved the use of HPCC, NAMD and Gridding and was the first trial to use HPCC instead of HPL.

HPCC A testing package for high performance computing that can evaluate the computing system for the speed of processor, memory access and provide useful information.

NAMD Used in biological science to simulate the large-scale molecule on parallel computers, it is able to scale from hundreds of cores to over 500,000 cores, and was the winner of the Golden Bell Prize in 2012.

Gridding As one of the key steps in the square kilometer array (SKA), it is also the most time-consuming part. Gridding refers to the mapping from the irregular data generated by the telescope to a regular 2D grid, and is a precondition of the Fourier operation. According to the rules, Gridding requires acceleration using MIC.

7.4.4 ASC Final: Close to Application

All teams were required to provide proposals that respected the 3 kW power budget. Besides the questions in the preliminary round, all teams need to finish HPL, WRL_Chem, PalaBos and a mystery question (HPCG) in the final, and then present their work. The questions in the final contain biology, mist, automobile design, and so on.

HPL evaluation is the key scenario for the TOP500 ranking, and was used to evaluate the teams' understanding of floating-point operations. A full study of the architecture, memory hierarchy, high-speed network, and algorithmic optimization, as well as the choice between homogeneous and heterogeneous architecture is needed to get a high mark. The committee encouraged teams to analyze the algorithm and code, and consider from a low level.

WRF-Chem is the one of the most advanced numerical prediction models that is widely used to predict environmental issues such as mist, PM2.5, PM10 and SO2.

PalaBOs is a fluent dynamic software that simulates the movement of objects and gas. It is widely used in the design of aircraft and automobiles and can be helpful to the resistance prediction.

HPCG is one of the three HPC scenarios.

7.4.5 A Gathering of Young Talent to Tackle Top Challenges

ASC15 encourages all teams to challenge the toughest problems in SKA, Gridding. With the purpose of building the largest synthesis radio telescope to solve questions and to help better understand the universe, SKA formations include countries like China and Australia. SKA needs to access data over 12 TB per second, and is almost 3.5 times larger than the outwards bandwidth of China Internet in late 2013. The big data sampled by SKA needs to be processed by a supercomputer, but the demand for computing power is more than the accumulation of the computing powers of all TOP500 supercomputers.

ASC15 choose the most complicated process of SKA—Gridding. Due to the fact that serial Gridding is far from satisfactory, SKA hope that through ASC15, a good solution could be found.

7.4.6 A Bonus of 250,000 for the Final

The ASC15 final was held in Taiyuan University of Technology, which was the first time it had been held in inland China. A total prize of 250,000 RMB was ready for the winners of: the Championship, the Runner-up, the ePrize Challenge, the Best Computing Performance prize, the Best Presentation and the Application Innovation Prize.

7.4.7 Zongci Liu "Presented" in the Closing Ceremony

In ASC15, the e Prize is awarded for Gridding optimization. Zongci Liu, who is a fiction writer and is famous for the book "Santi", sent a video congratulating the competitors. Santi gives a bold imagination of future universe, which will eventually be verified by SKA, with the help of supercomputing.

Chapter 8
Rules of ASC Student Supercomputer Challenge

8.1 Rules of the ASC14 Preliminary Stage

During the preliminary stage of ASC14, all registered teams are requested to submit a written proposal and the optimization results according to the tasks given by the organizing committee. The jury committee then determines which teams enter the final stage based on the submitted documents. In ASC14, only 16 of the over 100 participating teams got the chance to enter the final stage, which is very competitive. To make to the final stage requires a good understanding of the competition rules. Therefore, it is important to understand the regulations so as to be prepared for the finals. In this section, we will focus on the regulations of the preliminary round.

8.1.1 Brief Introduction of the Preliminary Stage

Dates: January 2, 2014 to February 27, 2014 (9:00am)

Format: The jury committee gave the specific tasks of the preliminary stage, which were sent to the participating teams by the organizing committee. Within the given time, the teams had to submit their written proposals and the optimization results according to the requirements. Through a careful evaluation of the submitted documents, the jury committee then decides which teams enter the final stage.

8.1.2 Detailed Rules of the Preliminary Stage

Each team has to submit its documents before the specified deadline to make it a valid submission. The submission should include all the items required by the organizing committee, such as the detailed proposal for the supercomputer system

ASC Community, *The Student Supercomputer Challenge Guide*,
https://doi.org/10.1007/978-981-10-3731-3_8

and the performance optimization results. The teams are prompted to read the description of the tasks carefully.

The proposal should include the following three parts:

8.1.2.1 A Brief Description of the University's or the Department's Supercomputer Activities (5 Points)

This should include but not be limited to:

(1) Supercomputing-related hardware and software platforms.
(2) Supercomputing-related courses, trainings, and interest groups.
(3) Supercomputing-related research and applications.
(4) A detailed description of the key achievements in supercomputing research (no more than two items), with proof materials attached (published papers, award certificates, etc.).

8.1.2.2 Introduction of the Team (5 Points)

Should include but not be limited to:

(1) Brief description about how your team is formed.
(2) Brief introduction of each team member (including group photos of the team).
(3) Your team slogan.

8.1.2.3 The Technical Proposal (90 Points)

This part is to evaluate the HPC knowledge of the team, and is also the most significant part of the proposal. The technical proposal not only evaluates the participators' knowledge about the hardware and system, but also makes every participator understand the importance and necessity of application and optimization.

The technical proposal should contain four aspects: (1) the proposed design of the supercomputer platform; (2) the HPL test; (3) the Quantum Espresso test; and (40 the CPU+MIC application optimization.

(1) Design of the supercomputer platform (15 points)

The whole system is requested to be within the 3000 W power budget. Therefore, it is necessary to find out the power of each component, for example, the CPU, memory, HCA card, and the hardware. Then it can be estimated how many machines can be deployed according to the power of each component, and decide whether to use accelerator cards or not according to the requirements of the applications. Finally, the system proposal should fully consider the overall situation and every specific detail. In order to achieve an excellent platform design proposal it is necessary to look deep into the hardware and system.

(2) The HPL test (15 points)

The goal is to get the highest FLOPS under the constraint that the accuracy must be validated.

The team is required to describe the operating system, complier, mathematical library, MPI, and so on, and provide the testing method, and to optimize the performance. The jury committee suggests a discussion at the algorithm level. This test is to evaluate the team for their understanding of the HPL algorithm and HPL model. Starting from the algorithm level would bring significantly more potential for the optimization part. With a good understanding of basic algorithm, code-level optimization becomes possible. Therefore, the content of this part should cover the HPL hardware/software environment, HPL basic algorithm analysis, HPL testing method, testing result, performance analysis, optimizing strategies, summary, and so on.

(3) The Quantum Espresso test (20 points)

The goal is to get the least running time as long as the accuracy is validated.

For this part, the team is also requested to describe the operating system, compiler, mathematical library, MPI, and so forth, the testing method, and the performance optimization. The method and procedure of software compiling should be described, and the key parameters listed. Unlike HPL, Quantum Espresso is the software from material science, and is used to solve realistic research problems. Therefore, to learn how to use the software is only the basic requirement for every team participating in the contest. It is more important to perform a deep analysis into the theory and the algorithm, which can be helpful for the testing and optimization afterwards. In this part, you should list the basic theory analysis, parallel method analysis, testing method, testing result, performance analysis, optimizing method, summary, and so on.

(4) CPU+MIC program design and optimization (40 points)

Under the constraint that the accuracy must be validated, you are required to parallelize the program given by the organizing committee either on CPUs or on CPU+MIC hybrid platforms. The organizing committee has provided the testing platforms for the participating teams, including both CPU clusters and CPU+MIC hybrid clusters. You may use different parallel programming methods, such as MPI, OpenMP, pThread, and OpenCL, to write the code, but all these methods should be supported by MIC. If the team chooses to use the MIC cards, the MIC cards must run in the offload mode. To perform the parallelization and optimization, it is suggested that participating teams learn about the physics model, the mathematical algorithm and general parallel computing techniques. Therefore, in this part, you should cover the introduction to the application, the physics model analysis, the mathematical method analysis, code analysis, vectorization analysis, design and implementation of the parallel program, testing, validation, and optimization strategies, and so on.

8.1.3 Summary of the Preliminary Stage

At the preliminary stage, each team is only requested to submit written documents. The jury committee has given a detailed specification about the required contents, and the points for different parts. The first few parts collect the basic information of the different teams. As long as all required items are covered, the teams should be able to get most of the points. The technical part accounts for 90% of the total score, including four different parts with different emphases. These tasks require the team to make comprehensive preparations on different aspects. The six members of the team should take responsibility for different parts of the tasks to take advantage of their different capabilities and expertise, and then combine the different components into one unified report.

8.2 Rules of the ASC14 Final Stage

After a careful evaluation of the proposals from over 100 universities around the world, the committee finally selected 16 teams for the final stage of ASC14.

8.2.1 Introduction to the Final Stage

Dates: April 21, 2014 to April 25, 2014

Format: within the 3000 W power budget, all teams should submit a proposal of the hardware configurations. The organizing committee will provide the different hardware components according to the proposal. Each team has two days to build the hardware platform, and to configure the operating system and other software components. The competition tasks (except for LINPACK) will be announced by the committee during the contest. Under limited power consumption, the team that finishes the running of the program within the shortest amount of time (given that the results are validated) will get the highest score. After finishing all of the tasks, each team will do a presentation to demonstrate their ideas, strategies, and optimization methods. The jury committee will combine the performance points and the presentation points to get the final score for each team.

8.2.2 *Detailed Rules of the Final Stage*

All teams should first learn about the exact time arrangements for the competition. They are requested to submit the performance results and other required items before the deadline of each task. Please read carefully the schedule of each task and the method for submitting the documents.

The final stage contains four major parts.

8.2.2.1 Building of the Hardware Platform, Operating System Installation, Parallel Environment Configuration, and Application Software Compilation

(a) During the preparation stage of April 21 and April 22, if you plan to amend the default hardware configuration, please inform the committee staff and keep the devices secure. The configuration should be restored to the default after the contest.

(b) The hardware configuration shall not be amended after April 23, and all machines in the cabinet should be turned on for the computation. If some devices break down, we provide backup machines as replacements. Note that the backup machines are shared among all teams.

(c) All teams should build the clusters at the scene. Each team is provided with a power supply PDU, which measures the power consumption of the system for the jury committee. During the entire period of the competition, the measured power cannot exceed 3000 W. If the power is over 3000 W for over 1 min, the results achieved during this period of time will be ruled as invalid, and the team should rerun the task. As the building of the hardware platform, OS installation, configuration of the parallel environment, and application software compilation are the bases of the competition, we suggest that each team should prepare and practice in advance.

8.2.2.2 The Rules for the Test of LINPACK, Quantum Espresso, SU2, LICOM, and a Mystery Application

(1) The items for each test and the related rules

- **LINPACK**: the LINPACK test will be held between 8:00 and 11:00, on April 23, which means the testing result should be submitted no later than 11:00. The result can be submitted ahead of time, but it cannot be updated once it is submitted (you can get the Quantum Espresso and SU2 test details once the LINPACK test is submitted).

- **Quantum Espresso**: we will provide the pseudopotential file, and the input parameter file. You cannot change them except for the two options that indicate the file location of the pseudopotential file and the output path of the specific temporary file.
- **SU2**: we will provide the *.su2(or *.cgns)file, and the *.cfg file. All test case files (*.su2, *.cgns, or *.cfg) cannot be amended.
- **LICOM**: we will provide all files required for the running of the model, such as the basic data files (INDEX.DATA, TSinitial, MODEL.FRC and dncoef.h1) and the namelist (ocn.parm). When required to continue a previous running of the model, we will provide the rpoiter.ocn file. The restart file (fort.22.yyyy-mm-dd) shall be obtained from the previous case. The first three test cases can directly use the licom2 binary for execution. For the other three cases, the teams are required to replace the param_mod.F90 file, and recompile. Except for the parameters that specify the path and the number of cores, no other parameters can be changed.
- **Mystery application**: we will announce the mystery application at 8:00 on April 24. The software, the test cases, the README file, and a piece of paper that contains the description of the rules will be provided to the teams. Please carefully read the README file before running and optimizing the mystery application.

(2) Run-time measurement and screen output

- All applications shall use the time command to record the real run time, and shall redirect the screen output to a result file. The result file will be used to validate the run time and the accuracy of the results.
- Example: $ time −p (your command here) > teamname.appname.workload No.time.txt 2>&1.
- The name of the result file must strictly follow the format specified by the organizing committee, otherwise the team will lose 0.5 points for each file.

(3) Code and document rules

- If the team amended the original code, the amended source code should be submitted using an USB flash disk.
- By the end of the competition, you need to submit a summary document, which should describe the test environment (OS version, MPI version, software version, etc.), the tuning operations (including the tuning targets and the methods), and the commands for running the programs, etc.

(4) Test cases and how to submit the results

See Table 8.1.

Table 8.1 Test cases and how results are submitted

Case	Input files provided in the USB drive	Upload to FTP	Results submitted through the USB drive
LINPACK	None	All the teamname.appname.workloadNo.time.txt files	teamname.appname.workloadNo.time.txt, README
Quantum Espresso	The pseudopotential file, the *.in file		teamname.appname.workloadNo.time.txt, README
SU^2	*.su2(or *.cngs), *.cfg		teamname.appname.workloadNo.time.txt, results, README
LICOM	INDEX.DATA, TSinitial, MODEL.FR, dncoef.h1, namelist		fort.22.yyyy-mm-dd (the last one for each case), teamname.appname.workloadNo.time.txt, README

8.2.2.3 Rules of the 3D_EW Task

(1) Date

- Regular session: 8:00–18:00, April 23, 2014.
- Final session: 8:00–14:00, April 24, 2014.

(2) Hardware rules

- Regular session: access to Tianhe-2 will be provided, and the performance test will take 128 nodes of Tianhe-2. Each node contains 2 CPUs (Intel(R) Xeon E5-2692 v2, 12 cores, 2.20 GHz) and 3 MIC cards (Intel(R) Xeon Phi 31S1P, 57 cores, 1.1 GHz, 8 GB memory). Each team uses 128 nodes exclusively in the Tianhe-2.
- Final session: The performance test will be running on 1024 nodes in Tianhe-2. Each node contains 2 CPUs (Intel(R) Xeon E5-2692 v2, 12 cores, 2.20 GHz) and 3 MIC cards (Intel(R) Xeon Phi 31S1P, 57 cores, 1.1 GHz, 8 GB memory). Each team uses 1024 nodes exclusively in the Tianhe-2.

(3) Procedure

- **Application debugging (April 21–April 22, 2014)**

(a) We will provide the account information of TH2 for the regular session at 8:00.
(b) We will allocate 128 nodes for each team.

- **Regular session (April 23, 2014)**

(a) We will announce the test cases at 8:00.
(b) Submit the result before 18:00. The TH2 nodes will be unavailable for use after 18:00.

(c) All teams should copy the optimized code, the makefile, the output (record.dat), the log file to their own user account: ~/3d_ew_out/, and download the optimized code, the makefile, and the log file to the USB drive. The USB drive should then be handed over to the committee staff (record.dat file should be put to the user's directory on TH2 only, no need to copy to the USB drive).
(d) As for the time measurement, all teams are requested to use time command to record the real run time and output real run time to the Time.txt file in the same directory, for example, time −p (your command here) 2> time.txt. The recorded time is the final score. The time file should be submitted together with the files mentioned in (c). (copy to ~/3d_ew_out/, and download to the USB drive).
(e) The rules for the file names:

Workload	outfile	Logfile	timefile	jobid
para11.in	record11.dat	log11.txt	time11.txt	
para12.in	record12.dat	log12.txt	time12.txt	
para13.in	record13.dat	log13.txt	time13.txt	

Note Please also submit the corresponding jobid, which will be used to count the run time of the task

- **Final session (April 24, 2014)**

(a) We will publish the TH2 account and the test cases at 8:00.
(b) Please submit the result before 14:00. TH2 nodes will be unavailable for use after 14:00.
(c) All teams should copy the optimized code, the makefile, the output (record.dat), and the log file to their own home directory: ~/3d_ew_out/, and download the optimized code, the makefile, and the log file to the USB drive. The USB drive should then be handed over to the committee staff (record.dat file should be put to the user's directory on TH2 only, no need to copy to the USB drive).
(d) As for the time measurement, all teams are requested to use time command to record the real run time and output real run time to the Time.txt file in the same directory, e.g. time –p (your command here) 2> time.txt. The recorded time is the final score. The time file should be submitted together with the files mentioned in (c). (copy to ~/3d_ew_out/, and download to the USB drive).
(e) The rules for the specifying the file names:

Workload	outfile	Logfile	timefile	jobid
para21.in	record21.dat	log21.txt	time21.txt	

Note Please submit the jobid, which will be used to count the run time of the task

8.2.2.4 Presentation

(1) Time: 08:30–12:30, April 25, 2014.
(2) Location: Classroom Building of Sun Yat-sen University South Campus.
(3) Rules:

- The presentation sequence will be determined by the draw at 14:00, April 24, 2014.
- The presentation should not exceed ten minutes otherwise the jury committee might downgrade the score accordingly. Each team can send two members and use PPT to make the presentation. Please use English for the oral presentation and in the PPT. After the presentation, there are three to five minutes for questions. Judges will give the final score (ten points at most) based on the performance of the presentation.
- During the presentation, only the mentor and the members of the same team can be present. Mentors and members from other teams should not enter the room.

8.2.3 Summary of the Final Stage

In the final stage, each team first needs to finish the building of the hardware system, and the system configuration within the given time. They should also finish the testing and the optimizations of applications within the given time and within the 3000 W power budget. Therefore, it is necessary to do some practice in advance to get familiar with the hardware and the system. In terms of the application preparation, each team should become familiar with related applications, build a rough concept about the relationship between the run time and the input files, and perform optimizations according to the characteristics of the applications. For the 3D_EW task, the usage of MIC is required. Therefore, it is important to look deeply into both the code and the MIC architecture so as to make the most usage of the MIC computing power. Finally, the presentation should be well structured, providing an interesting and clear demonstration of the testing and optimization efforts.

Part III
Advances

Chapter 9
Competition Proposal

9.1 Introduction

The ASC student cluster preliminary competition requires a proposal submission as a perquisite for evaluation by the competition judges for finalist entry selection. This chapter will provide some insights to the efforts and suggested methods required for preparing the proposal and guiding a team to overcome these challenges at a first attempt. The proposal demonstrates the students' theoretical knowledge on cluster competition subjects ranging from hardware architecture, parallel programming, and software applications, to code optimization. This preliminary stage also provides an excellent chance to learn practical aspects of high performance computing (HPC) or supercomputing in a very competitive learning environment.

During the training phases, students learn valuable skills that will help them in their future working careers, such as team skills, project planning skills, and presentation and technical writing skills. They also have the opportunity to interact with experts during the various courses and Q&A sessions.

This guide serves to prepare future aspiring participants on technical writing, and to provide advice about the best practice and strategies in preparing and submitting a professional document. The road to the finals begins with this initial task of writing this competition proposal, which will be a valuable experience for your undergraduate life. The following chapters provide the necessary details on how to prepare and achieve this target.

ASC Community, *The Student Supercomputer Challenge Guide*,
https://doi.org/10.1007/978-981-10-3731-3_9

9.2 Team Selection and Composition

The primary task for the students who are interested about this competition is to form a team with members that have various skill sets, to undertake the various challenges outlined in the proposal.

The ASC competition will require substantially more preparation efforts when compared to International Supercomputing Conference (ISC) and Supercomputing Conference (SC) competitions. However, the efforts preparing the proposal provide the students with a better insight into the world of supercomputing. The advisor of team would also provide help on forming the team and preparing the proposal.

9.3 Basic Requisite

It may seem difficult at first glance to prepare for this daunting task. The guidance will provide an overview to aid participants to overcome this challenge. As a basic requirement, basic Linux knowledge and skills, and experience using HPC servers would be highly beneficial. Students may have their first exposure to HPC during their academic studies or through their internships in the industry (if applicable).

It is recommended that participation for the competition receives official support from the university, and a dedicated mentor can be appointed to support and train the students. Note that for a team of students with great capabilities and experience on HPC, they might also be able to accomplish the task of writing the proposal by themselves.

9.4 Training

Training forms an essential component and foundation to provide participants with the basic core concepts to prepare the competition proposal. As the timeline is rather short and spread over Christmas, New Year's Day, and the Chinese Spring Festival, we suggest an extremely detailed training plan. A well-developed training program should be established to cover subjects including computing theory, parallel programming, computer hardware, mathematic, and software applications. As part of the training curriculum, short work assignments can be incorporated to complement the process of learning and to promote student interest. Students with other engineering backgrounds rather than computer science might find it difficult to learn these subjects at the beginning. However, their experience on the actual scientific or engineering problems can be quite helpful in the process of optimizing the application software.

9.4.1 *Training Plan*

The training can be developed in two phases. The first phase is basic theory during the initial stage prior to the release of the preliminary competition details and requirements. Such training will usually be classroom based. Second phase training can be targeted more specifically towards working on the required tasks in the proposal. This includes hands-on practice. (Hint: Participants may refer to the previous years' competition requirements for an overview of the expected tasks, usually including design of the hardware platform, LINPACK test, test, and optimization of other Scientific/Engineering applications.)

As most students may/do not have any prior knowledge of working within a supercomputing environment, it is imperative that the training materials are written in simple language and terms to facilitate their understanding, interest, and aid their progressive understanding to more advanced contents.

Providing essential training support through offline consultation is very helpful in providing students the opportunity to clarify their doubts and concerns on a personal basis. (It is recommended the trainer also prepare the subject matter in advance as part of the training effort before providing consultation.)

To ensure a complete range of training content is provided to the students it is essential to collaborate between your university and a research institute or industry vendor or other schools in your university. As there are currently three established competitions (SC, ISC, and ASC) across the world today, there is also considerable online information available for training preparation.

9.5 Training Hardware

The competition proposal provides a balanced content covering design and practice-oriented tasks where participants are required to progressively learn and put their skills and knowledge to the tasks of testing and optimizing scientific/engineering applications. It is beneficial to seek support from a hardware vendor for the hardware platforms used for training. Leveraging on existing vendors' partnerships can be a great starting point. On the other hand, any additional server equipment from your research center could also be utilized to setup and establish a mini cluster for hands-on practice.

9.5.1 *Hardware Platform*

There are several types of hardware test platforms available for this purpose: laptops where participants may practice operating system installation and application compilation, a two-node cluster for testing and code compilation for parallel programs (MPI). This two-node cluster is extremely helpful in ensuring that repeatable results can also be replicated on the test platform provided by the ASC competition organizer. It would be ideal if students could make use of this short window to also learn to setup a 2-node cluster as part of the program.

Variations in performance run results are expected when porting codes across various test platforms. Therefore it is essential to understand the cluster setup, software stack, and interconnect specification and configuration.

9.6 Competition Proposal

The requirements of the proposal will be made available to participants during the formal release of the competition details. The proposal serves as a formal report designed to convey technical information in a clear and easy-to-understand style. It will be divided into chapters, which allow different readers to access different levels of information.

9.6.1 Proposal Sections

It is recommended that the proposal is prepared with separate sections: a main section for key findings and appendices for supplementary information. Concepts and result analysis trends should be represented in diagrams and graphs as they provide an alternate form of presentation to the readers/judges to complement your proposal.

9.6.2 Guidelines for Graphs and Diagrams Presentation

The following section discusses the suggested guidelines for presenting graphs and diagrams to support your findings in a proposal. A graph is very useful for visualizing and describing the relationship between two variables (i.e. speedups vs. CPU cores). The independent variable (configurable variable, i.e. CPU cores) is plotted on the x-axis and dependent variable is plotted on the y-axis (responding variable, wall time). Both axes of the graphs should also be labelled with both quantity and units. The graph scales should also fill the entire page if possible as shown in Fig. 9.1.

Illustration with diagrams using a suitable font makes the diagram presentable. Objects and fonts are interrelated when they are used to convey a subject/topic. Both the font type and objects in a diagram translate and visualize an idea or concept. Usage of lines in diagrams should be consistent. If a dashed line has been used to indicate Ethernet connection, do not use the same line pattern for InfiniBand connection in your cluster.

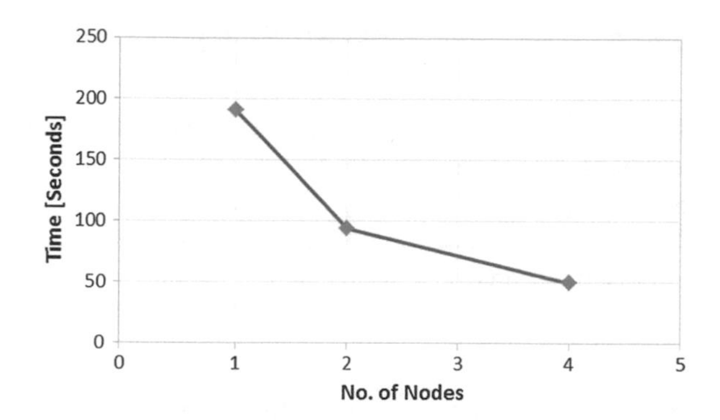

Fig. 9.1 Workload performance result

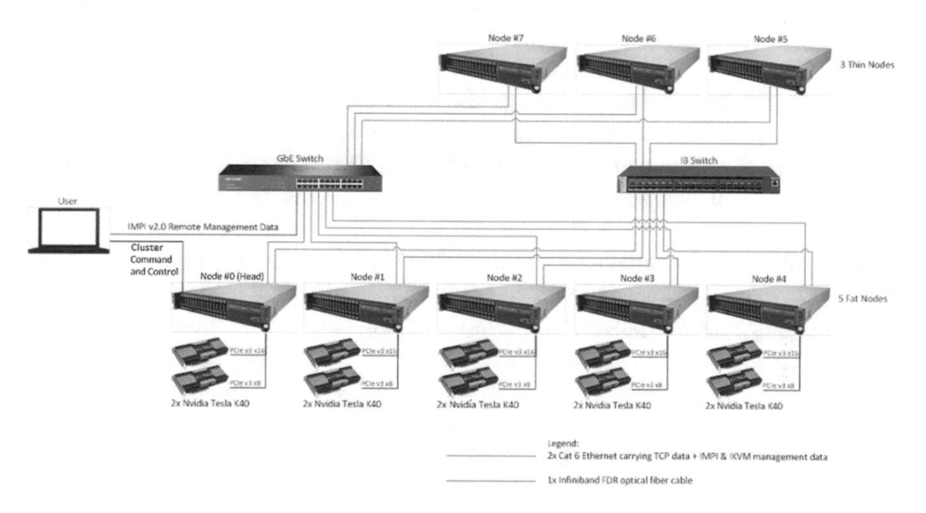

Fig. 9.2 Example of figure with whitespace and line variation

A balance between the objects and whitespace is essential in creating a good diagram. Whitespace can help to emphasize particular elements but can also help to balance the objects in the diagram. It must be noted that using black-space is not highly recommended as it adds a visual strain to the reader's eye and the font colors of the text will not be easily readable. Shadows give the diagram an unclean and artistic feel; therefore its use is not recommended. Example of object layout and varied linear are shown in Fig. 9.2.

9.6.3 Guidelines for Other Sections

Acknowledgments may be made to individuals or institutions that are not mentioned in the work but have made significant and important contributions.

9.6.4 Knowledge of High Performance Computing Activities

This section is a relatively simple section that requires you to describe HPC activities in your university or institute. You may provide a description of the hardware platform and the type of research activities based on the platform. It will be helpful if you can also provide information of the scale of the massively parallel jobs being executed on the system.

9.6.5 Hardware Design and Energy Efficiency

The hardware system design section is not directly linked to the domain software applications. Therefore you do not need to establish a similar hardware setup to run the jobs based on your theoretical design. In this section, it is important that you understand the importance of having the essential compiler and math libraries installed on your test cluster. Having access to a small cluster will provide participants with the chance to verify their findings. These libraries form the basis for ensuring that you can successfully run your LINPACK test and quantify your findings.

You may test and run them on any available hardware cluster you have access to. It is also highly recommended that you consider discussing the effects of hardware tuning for keeping the power consumption within the 3 kW constraint. Prior to having access to an operating cluster, tweaking the bios of the server and attaching a power meter to the wall socket provides valuable information on the server's energy consumption performance. Table 9.1 shows the summary of power consumption in watts and the energy efficiency details.

Servers are generally equipped with redundant components for reliable operations in critical business environments. We can consider using energy efficient devices or novel cooling techniques (water cooling) for reducing the power consumption of the cluster. The power usage of powering your water pumps for water cooling needs to be covered within the 3 kW constraint. (Note: You might need to

Table 9.1 Test results with different number of CPU cores, with the configuration of a single node with a single GPU accelerator

Cores	Accelerators	Resulting Rmax [GFlops]	Power consumption [W]	Energy efficiency [GFlops/W]
24	1 K40 GPU	1440	700	2.057
18	1 K40 GPU	1364	635 Est.	2.148
12	1 K40 GPU	1253	570	2.198
6	1 K40 GPU	1116	500	2.232
4	1 K40 GPU	1046	475	2.202
2	1 K40 GPU	895	465	1.924

Table 9.2 Energy efficiency of different architectures

Nodes	CPU cores	Accelerator	Parallelization technique	N/NB	P × Q	Rmax [GFlops]	Power [W]	Energy eff. [Gflops/W]
1	24	None	OpenMP within node	87,000/ 224	1 × 1	473.3	450[a]	1.051
1	24	1 Xeon Phi	OpenMP within node, offload to Xeon Phi	87,000/ 1024	1 × 1	1185.6	700[a]	1.693
1	24	1 K40 GPU	OpenMP within node, CUDA to GPU	87,000/ 1024	1 × 1	1440	700 Est.	2.057
2	48	1 Xeon Phi per node	OpenMP within node, MPI across nodes, offload to Xeon Phi	123,000/ 1024	1 × 2	2216.9	1400 Est.	1.583
2	48	1 K40 GPU per node	OpenMP within node, MPI across nodes, CUDA to GPU	123,000/ 1024	1 × 2	2471	1400 Est.	1.765

Notes: *CPU* Intel Xeon E5-2695v2: 2.4 GHz * 12 cores * 2 socket * 8 Flops/clock = 460.8 GFlops per node; *MIC* Intel Xeon Phi 5110P: 1.053 GHz * 60 cores * 16 Flops/clock = 1010.88 GFlops per accelerator; *GPU* NVIDIA Tesla K40: according to specifications = 1430 GFlops per accelerator; [a]Power was measured at the outlet with a power meter sacrifice the total number of CPU cores to attain the best Floating Point Operations per Second (Flops) in LINPACK testing.)

This section will demonstrate your understanding of hardware energy efficiency. All final optimized parameters and the necessary accompanying justifications should be documented as shown in Table 9.2.

9.6.6 High Performance LINPACK (HPL)

High Performance LINPACK (HPL) is a computation benchmark that measures the performance of an HPC system by solving a dense matrix of linear equations. In this particular section, understanding the theory and algorithm concept will be helpful in providing the opportunity to evaluate and tune the parameters to achieve your desired results. It will be helpful to know that different versions of the HPL code will provide significantly different results. Therefore it is recommended that you adopt a consistent version number and libraries across different platforms during your tests.

The findings from this section of the HPL test and tuning should be documented and justified to support your selected parameters. Table 9.2 shows an illustration of the findings summarized in table format which you may adopt.

9.6.7 Application Software

The first step in this section is to install the software from the source code. There are accompanying instruction files in the source file that you have downloaded. Take a moment to read the contents. Alternatively the developer's website may also contain valuable information and instructions to proceed with your code compilation. It is easier to start with compiling the software with a serial configuration, running the test, and achieving a valid output. This forms a baseline for your verification with parallel runs. Installing the code with a parallel configuration may require more efforts for a first-time user and this will improve with more practice.

During the performance optimization process, you may compile the code with combinations of different compliers and MPI libraries. This is generally an iterative process but you will gain a lot after this challenging process. You may refer to HPC advisory council's website for best practice information on installing various scientific and engineering codes. You may start off by installing the code by following the suggested options and gain confidence in code compilation.

9.6.8 *Configurations and Assumptions*

For the tests of domain applications, it is recommended you document your understanding of the basic theory and the real-world application. You can then run performance tests to verify and enhance your understanding of the code. In fact, understanding the application characteristics is more important than having the best hardware. If the software is not able to leverage the latest hardware architecture, you will not be able to achieve the maximum performance. Amendments to source codes and input files should be documented in your proposal to compare with the baseline sample code. The following are some suggested approaches:

9.6.8.1 Installing Code with Recommended Parameters

Based on those materials provided on Quantum Espresso webpage and in order to make Quantum Espresso run at its best efficiency, it must be configured with all correct settings. The following setup shown in Table 9.3 was used as a starting point for running the test.

9.6.8.2 Identifying Parameters that Affect the Application Performance

The following parameters have been identified to potentially increase the performance of Quantum Espresso. Specific values used for testing are placed in the parentheses.

- OpenMP Threads (1, 2, 3, 4);
- total number of cores, i.e. OpenMPI cores * OpenMP Threads (12, 24, 48);
- npool; and
- Ndiag.

For OpenMP threads, a small number of threads is recommended by the National Energy Research Scientific Computing Centre. As OpenMP works differently from MPI, we generally take an MPI-OpenMP hybrid parallelization scheme to improve parallel performance.

Table 9.3 Baseline setup values

Configuration	Value
Usage setup	Enabled OpenMP, enabled scalapack
Compilers	Compiled with C/C++—Intel C/C++ (icc), Fortran 77/90—Intel Fortran (ifort) and Intel MPI (mpiicc and mpiifort)
Libraries	Built with MKL BLAS, LAPACK, SCALAPACK, BLACS and FFT

9.6.8.3 Focusing on the Configurations Related to Our Specific Problem

Quantum Espresso is written as highly scalable software with a parallelization scheme of five levels: image, pool, plane-wave, task group, and linear-algebra parallelization. However, in our case, we only use two parallelization options, which is pool (distribute k-points among npool of CPUs) and linear-algebra parallelization (distribute and parallelize matrix diagonalization and matrix-matrix multiplications). The rest of levels were not tweaked because they are not related to our problems, that is, image and plane-wave parallelization, and the system is not big enough for such a scale of parallelization to take place.

9.6.8.4 Reviewing the Run Type Specified in the Input File

From the input files given, we can see that workload 1 is about structure relaxation and workload 2 is just a simple self-consistency run.

Judging from the position of atoms, which is visualized in Fig. 9.3, workload 2 is actually dealing with an intrinsic defect, O-vacancy, in a zirconium oxide supercell. Since this kind of intrinsic defects often introduce rich luminescent properties, understanding these defect-related excitations is important to physicists or material scientists and they are particularly important for design and optimization of some nano-materials.

In fact, the calculation was performed in the reciprocal space that deals with the Brilloin-zone with ease. In general, the K-point mesh method is used to spread equally in the Brilloin-zone, which is known as the Monkhorst and Pack method. In workload 1, four K-points were implemented. Therefore along each reciprocal lattice-vector, four points were considered, while in workload 2, only the center point was considered.

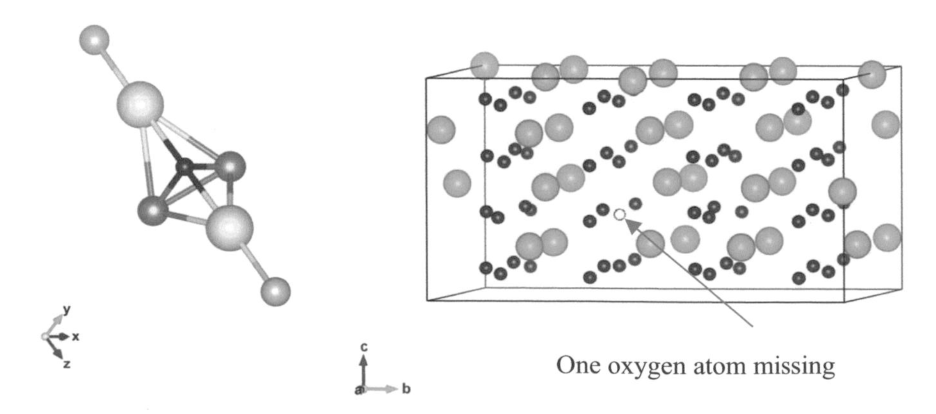

Fig. 9.3 Structure of workload 1 and workload 2

In summary, the number of atoms along with parameters encut, nspin, con_thr, kpoints will largely determine the performance and the run time.

9.6.8.5 Discussion and Review of Results

npool is a factor that reduces communication by agglomerating the workload. In the above data, it shows that time decreased from npool 1 to npool 2, which is mainly because we have more computing resources to process the problem. After that, increasing the number of nodes would generate more communication, leading to a longer run time. The highlighted red bar in Fig. 9.4 shows a possible overhead of generating the pool, which shows a linear increment (10, 20, 30, 40 s respectively for an increasing npool from 1 to 4).

The fluctuation of the results from Fig. 9.5 was not explainable by any existing theories. Thus, we came up with two plausible reasons.

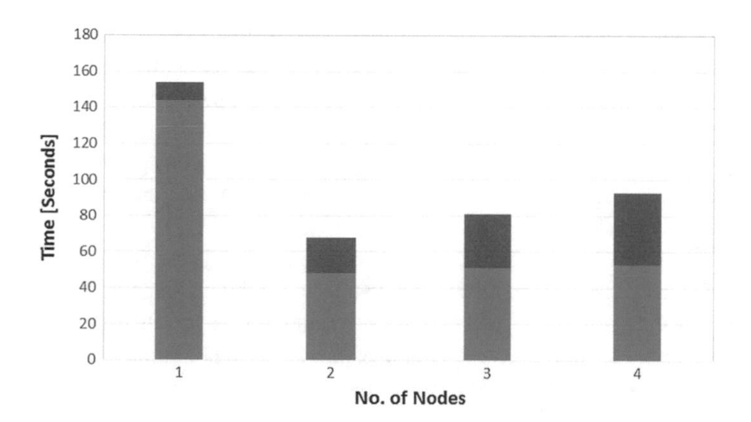

Fig. 9.4 Workload 1 npool comparison

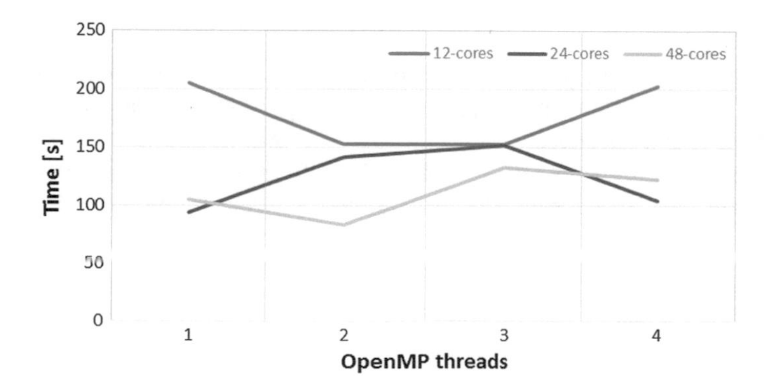

Fig. 9.5 Time taken by different OpenMP threads with different running cores

- **Conflicting hardware access**: The difference between OpenMP and MPI is that OpenMP threads work on the same set of memories and may result in conflicts when both threads enter heavy data access phases.
- **Untuned MPI**: During the server setup, Intel MPI Automatic Tuning Utility was not used to check if MPI communication produces consistent speed.

9.6.9 Code Optimization

For code optimization, there will be a challenge for both the computer science and application domain (scientific or engineering) experts. The code optimization section is assigned with the highest score.

The optimization can be attempted from either the programming perspective or the algorithm perspective. It must be noted that rewriting the mathematical algorithm requires extensive efforts.

For the optimization of the code, you may like to first run the correctness test to validate the generated output. Achieving a good performance for HPC applications involves a good understanding of the workload through performing profile analysis, and comparing behaviors of using different type of hardware architecture (i.e. CPU vs. CPU + MIC). This process will pinpoint the bottlenecks of the program. Working on the largest bottleneck is an easier option to tune and improve the application performance.

After the optimization, you need to document your performance optimization techniques, including your modification to the compilation flags and configuration parameters of the program. It is highly recommended that you display code snippets (before and after) to illustrate the improvements. The performance benefits from different optimization techniques should also be documented for comparisons.

9.7 Task Distribution

It is recommended that you split up the tasks among different team members to work on different sections of the proposal. This "parallel" approach is likely to be more efficient and effective.

Among the team members, the leader will perform two roles, a member to work on a specific task and a leader to motivate and coordinate the entire team. Time spent on various task varies accordingly to the efforts that the team wish to go to. This ranges from 10 to 200 h spent to finish each task.

9.8 Summary

Attempting to prepare and submit such a proposal requires a team effort. It will be an initially daunting task but having a proper training framework in place will be highly effective and beneficial. Having a group of highly motivated individuals will form the basis from which to establish a team that prepares and submits the proposal. The proposal incorporates a substantial hands-on effort to work on the application software testing and performance optimization, therefore having a strong foundation of the related skills and knowledge is very important.

The suggestions and guidelines presented in this chapter will provide first-time participants suitable guidance and advice to prepare a plan and the team. Presenting the proposal in a suitable format provides a consistent platform for evaluation. Technical concepts and findings should be correctly documented and justified. In short, the following suggestions will form the basis for a good proposal:

- good basic understanding of the fundamental theory;
- teamwork and co-ordination;
- good time management;
- self-learning; and
- progressive review with advisor and other experts on proposal progress.

The proposal will demonstrate the team's basic understanding of the concepts and capabilities to test and optimize application software. Finally with all essential findings documented, verified and proofread, the proposal will be submitted to the jury committee for evaluation and judging.

Chapter 10
Design and Construction of the Clusters for the Competition

The history of cluster use can be dated to the middle 1990s. In recent years, cluster-based architecture has become more and more popular due to its cheap price, good flexibility, and good scalability. There are more and more cluster-based supercomputers in the Top500 list, since the first cluster entered the list in June 1997. In November 2003, there are already 208 cluster-based supercomputers in the Top500 list. In the recent 23 publications of the Top500, the top supercomputers are all based on cluster architecture. For example, the current top supercomputer is the Tianhe-2 supercomputer in Guangzhou China, designed and developed by the College of Computer, National University of Defense Technology. Therefore, the cluster is definitely the most successful system architecture in the field of high-performance computing.

In the ASC contest, each team is requested to build a cluster within the 3000 W power budget. In this chapter, we will briefly introduce the cluster architecture and how to build it.

10.1 Architecture of Cluster

As an economical, easily constructive and scalable computing system for parallel computing, the cluster contains homogeneous or heterogeneous compute nodes connected by a high-performance networking system. From the perspective of users, clusters can be treated as a single computing resource. Figure 10.1 shows the general architecture of a PC cluster that contains 4 PCs being connected by a Switch. NIC refers to the network interface, and PCI refers to the I/O bus. It is a shared nothing structure, which is widely used for most clusters. If we replace the Switch with a shared memory, it turns into a shared-memory cluster structure.

Each computer in a cluster is called a node. Each node itself is a complete system with a local disk and operation system, and can supply computing resources to

ASC Community, *The Student Supercomputer Challenge Guide*,
https://doi.org/10.1007/978-981-10-3731-3_10

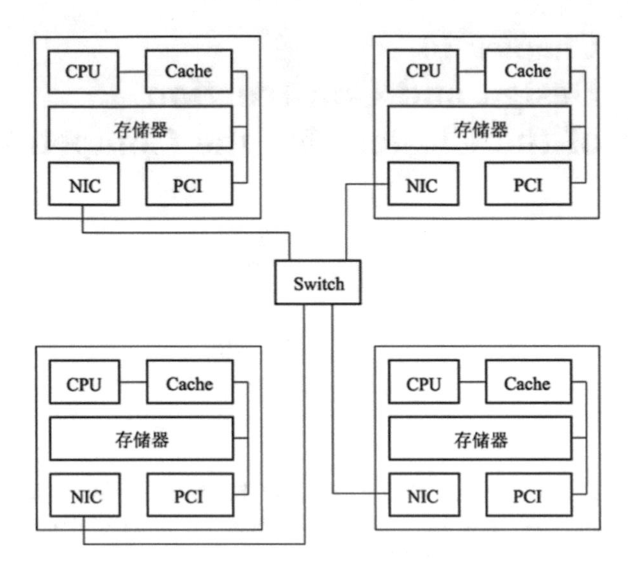

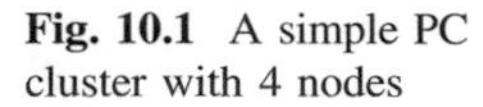
Fig. 10.1 A simple PC cluster with 4 nodes

users. In addition to the PC, a workstation or even a more powerful symmetric multi-processor can also be used to form a node.

The networking system connecting each node together generally uses commercial network such as Ethernet, Myrinet, InfiniBand, and Quadrics. Some networking systems choose special networks such as SP Switch, NUMAAlink, Crossbar, or Cray Interconnect. The networking interface and the I/O bus of each node are connected in a loosely coupled way, as shown in Fig. 10.1.

ASC mainly uses InfiniBand and Gigabit Ethernet as the network connection. Inspur provides a networking environment with complete InfiniBand and Gigabit Ethernet systems.

10.2 Performance and Power Consumption of Cluster Systems

Based on the number of nodes and the configuration of each node, it is easy to estimate the peak performance of the entire cluster system. For example, supposing we have a cluster with 8 nodes, and each node contains 2 Intel Xeon E5 2692v2 CPU (12-core) and 3 Intel Xeon Phi (57-core), the peak performance for each node would be:

- CPU: 12 Core $\times$ 8 Flop $\times$ 2.2 GHz = 211.2 GFlops
- Phi: 57 Cores $\times$ 4 Threads $\times$ 4 Flop $\times$ 1.1 GHz = 1.003 TFlops
- Node: 0.2112 $\times$ 2 + 1.003 $\times$ 3 = 3.4314 TFlops
- Cluster: 3.4314 $\times$ 8 = 27.4512 TFlops

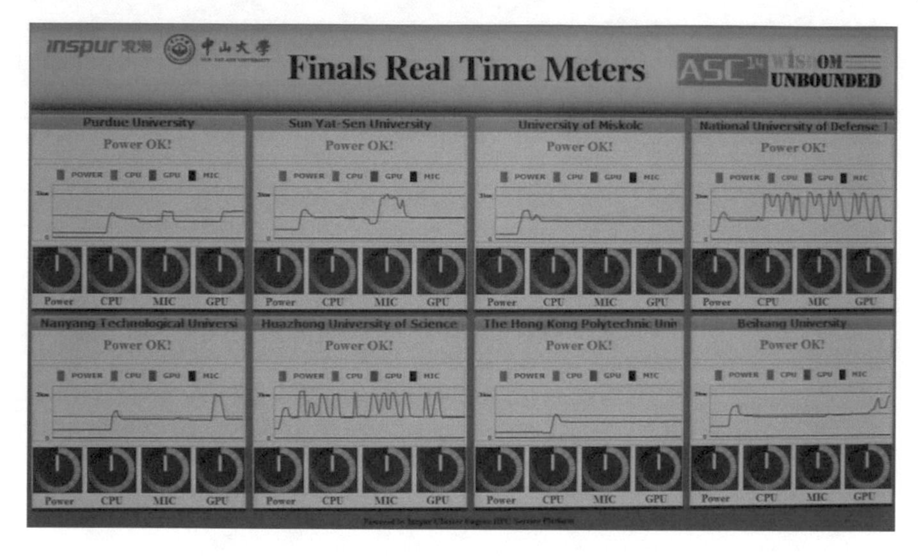

Fig. 10.2 The measurements of the real-time power consumption from some teams in ASC14

Likewise, it is easy to estimate the overall power consumption of the cluster by adding up the power consumption of each component. Under the allowed maximum power consumption, you can decide the number of nodes to use.

Some teams may try to use low-power components to decrease the overall power consumption. For example, they may use SSD to replace HDD. Even though such behavior cannot save enough power for adding another node, it has many benefits because the committee usually uses a current meter to monitor the real-time power consumption.

Figure 10.2 shows the measurements of the real-time power consumption from some teams in ASC14. Obviously, using SSD could decrease the power consumption so that the system could have more space in terms of power consumption.

To increase the efficiency and save more time for the debug and testing, you can either use script to automate the system construction and configuration, or install the system on the SSD disk.

Chapter 11
Optimization for the High Performance LINPACK Benchmark

HPL (High Performance LINPACK) benchmark can reflect the system's capacity to do floating-point operations, and is the most popular way to evaluate the performance of the system. The two releases of the Top500 list every year use the HPL test to rank different systems.

The HPL benchmark is to solve the one-dimension and N-order dense linear algebraic equations through the Gaussian Elimination method. The number of floating-point operation needed when the size of matrix is N would be (Flop: Floating-point operation):

$$\left(2/3 \times N^3 - 2 \times N^2\right) \tag{11.1}$$

Therefore, the performance of the system that solves the N-size problem in time T would be: (Flops: Floating-point operation per second):

$$\text{Performance} = \text{Flop/T} = \left(2/3 \times N^3 - 2 \times N^2\right)/T \tag{11.2}$$

In the latest release of Top500 list, Tianhe-2 supercomputer ranks top for the fourth time, with a theoretical peak performance of 53.9 PFlops, and a sustainable performance of 33.9 PFlops. Under the restriction of the 3KW power budget, the performance in the HPL is usually around a few TFlops. In 2014, the performance managed to exceed 10 TFlops.

During the HPL test, the size of the matrix N can be changeable on demand. To obtain the maximum performance, any optimizations without changing the formula form are allowed. For example, you can adjust the value of N, change the number of CPU, or use the high-performance mathematical library.

ASC Community, *The Student Supercomputer Challenge Guide*,
https://doi.org/10.1007/978-981-10-3731-3_11

11.1 Brief Introduction of HPL

HPL can evaluate the Flops of the computer system by solving a dense linear equation $Ax = b$, where A and b are given, and x is the vector to solve.

HPL employs the Gaussian Elimination method. The basic idea is to first LU-decompose matrix A ($A = LU$), where L is a unit lower triangular matrix, and U is a non-singular upper triangular matrix. Let $Ux = y$, $Ly = b$, so that we are going to solve $Ux = y$.

More specifically, the decomposition of A:

$$A = \begin{pmatrix} A_{11} & A_{12} \\ A_{21} & A_{22} \end{pmatrix} = \begin{pmatrix} L_{11} & 0 \\ L_{21} & L_{22} \end{pmatrix} * \begin{pmatrix} U_{11} & U_{12} \\ 0 & U_{22} \end{pmatrix} = L * U \tag{11.3}$$

where A_{ij} and L_{ij} are given, and U_{ij} is what we need to solve.

Based on Eq. (11.3), we can figure out:

$$A_{11} = L_{11} * U_{11} \tag{11.4}$$

L_{11} and U_{11} in Eq. (11.4) are calculated through decomposing the row main elements, so that L_{21} and U_{12} can be calculated as follows:

$$L_{21} = A_{21} * U_{11}^{-1} \tag{11.5}$$

$$U_{12} = A_{12} * L_{11}^{-1} \tag{11.6}$$

In addition, we can also get the following equation through Eq. (11.3):

$$A_{22} - L_{21} * U_{12} = L_{22} * U_{22} \tag{11.7}$$

Based on Eqs. (11.4)–(11.7), we can figure out that the key computing kernel for HPL is the matrix calculations, including matrix multiplication/addition, and the inverse matrix multiplications.

11.2 Installation and Test

We introduce the installation and test of HPL using the Ubuntu 14.04 as the targeting operating system. The computer is a Lenove X1 Carbon with 2 Intel Core i7 dual-core CPUs, 8 GB memory and the gfortran compiler.

11.2.1 Compiling ATLAS

By default, ATLAS is required for HPL. After decompressing the ATLAS package, you can access a directory named ATLAS. Compile ATLAS based on the following steps,

- *cd ATLAS*
- *mkdir build*
- *cd build*
- *../configure*
- *make*

If successfully compiled, libraries like libatlas.a, libcblas.a can be found in the build/lib directory and are required for compiling the HPL.

11.2.2 Compiling HPL

Download the HPL package (e.g. version 2.1) from the website, and decompress it, then you can found a directory named hpl-2.1.

You need to create a Makefile before compiling the HPL. Based on the naming convention of HPL, that is, Make.arch, the name of the Makefile should be Make. Ubuntu.

You can get the Makefile by revising the Make.* file in the setup sub-directory. As the Intel processor is used, you can directly build makefile based on the file of setup/Make.Linux_PII.CBLAS_gm. Required revisions include:

```
- ARCH = Ubuntu
- TOPdir = ...               # TOPdir refers to the directory of
  hpl-2.1
- LAdir = ...                # LAdir refers to the location of
  libatlas.lib
- LAinc = ...
- LAlib = $(LAdir)/libcblas.a $(LAdir)/libatlas.a
```

Compile the HPL using the following command:

- *make arch=Ubuntu*

After a successful compile, two new files will be generated in/bin/Ubuntu, namely the HPL.dat, and xhpl. HPL.dat is the configuration file that records some parameters for configuration, while xhpl is the executive program.

11.2.3 Configuration Modification

The following part shows an example of HPL.dat, the parameter of which is listed in Table 11.1.

In the configuration file, on the left-hand side of each row is the value of each parameter, while on the right-hand side lists the actual meaning. Some parameters should be used as a group. For example, # *of problems sizes (N)* in the third row refers to the number of sizes to be used in the test, while the number in the 4th line indicates the specific size value (*N* = *8192*). There will be one test and one result for each different value of *N*.

Table 11.1 HPL.dat

Row	HPLINPACK benchmark input file	
1	*HPL.out*	*output file name (if any)*
2	6	*device out (6=stdout,7=stderr,file)*
3	*1*	*# of problems sizes (N)*
4	*8192*	*Ns*
5	*1*	*# of NBs*
6	*192*	*NBs*
7	*0*	*PMAP process mapping (0=Row-,1=Column-major)*
8	*1*	*# of process grids (P x Q)*
9	*2 1 4*	*Ps*
10	*2 4 1*	*Qs*
11	*16*	*threshold*
12	*1*	*# of panel fact*
13	*0 1 2*	*PFACTs (0=left, 1=Crout, 2=Right)*
14	*1*	*# of recursive stopping criterium*
15	*2 4*	*NBMINs (>= 1)*
16	*1*	*# of panels in recursion*
17	2	*NDIVs*
18	*1*	*# of recursive panel fact*
19	*0 1 2*	*RFACTs (0=left, 1=Crout, 2=Right)*
20	*1*	*# of broadcast*
21	*0*	*BCASTs (0=1rg, 1=1rM, 2=2rg, 3=2rM, 4=Lng, 5=LnM)*
22	*1*	*# of lookahead depth*
23	*0*	*DEPTHs (>=0)*
24	2	*SWAP (0=bin-exch, 1=long, 2=mix)*
25	64	*swapping threshold*
26	*0*	*L1 in (0=transposed,1=no-transposed) form*
27	*0*	*U in (0=transposed,1=no-transposed) form*
28	*1*	*Equilibration (0=no,1=yes)*
29	8	*memory alignment in double (> 0)*

For example, based on the following setup of HPL.dat, there will be two tests in total with the size of the matrix to be 8192 and 4096, respectively. The space is used to isolate different parameters.

```
- 2              # of problems sizes (N)
- 8192 4096 Ns
```

Similarly, row 5 and row 6 indicate the size of the blocking. Row 8, 9, and 10 indicate the array of processes, and so on. Therefore, one configuration of HPL.dat can lead to different tests. The total number of tests equals the product of each parameter. For example, the following configuration can lead to 6 tests.

```
- 2              # of problems sizes (N)
- 8192 4096    Ns
- 3              # of NBs
- 192 96 256   NBs
```

11.2.4 HPL Testing

The following command can be used to run the HPL:

```
- mpiexec -n 4 ./xhpl
```

Figure 11.1 shows the screen output, in which the content of each parameter is omitted.

Therefore, the performance of the test should be 1.197e+01 GFlops, that is, 11.97 GFlops.

11.3 General Optimizations for HPL

11.3.1 Modifying the Configuration Parameters

The value of the parameters defined in hpl.dat has a big influence on the performance, especially the size of scale N, size of block NB, and the array of processes $P \times Q$. It is easy to determine optimal values of those parameters according to the configuration. The influence of other parameters, such as the array arrangement of

```
… ….
--------------------------------------------------------------------------------
- The matrix A is randomly generated for each test.
- The following scaled residual check will be computed:
      ||Ax-b||_oo / ( eps * ( || x ||_oo * || A ||_oo + || b ||_oo ) * N )
- The relative machine precision (eps) is taken to be              1.110223e-16
- Computational tests pass if scaled residuals are less than          16.0
================================================================================
T/V                N    NB    P    Q               Time                 Gflops
--------------------------------------------------------------------------------
WR00L2L2        8192   192    2    2              30.63              1.197e+01
HPL_pdgesv() start time Sat Dec 13 01:55:16 2014
HPL_pdgesv() end time   Sat Dec 13 01:55:47 2014
--------------------------------------------------------------------------------
||Ax-b||_oo/(eps*(||A||_oo*||x||_oo+||b||_oo)*N)=        0.0023562 ...... PASSED
Finished      1 tests with the following results:
              1 tests completed and passed residual checks,
              0 tests completed and failed residual checks,
              0 tests skipped because of illegal input values.
--------------------------------------------------------------------------------
End of Tests.
================================================================================
```

Fig. 11.1 Part of the screen output of one HPL test

the processor, the matrix decomposition method, transverse broadcasting method and so on, is relatively small. Besides, it is also difficult to determine the optimal values of those parameters in single test.

11.3.1.1 Size of Matrix

In general, the larger the size of the matrix, the higher the Flops performance, as the proportion of effective computation is larger. However, the increase of the matrix size may exceed the cache capacity, which in return leads to the performance decreases. Therefore, it is necessary to carefully decide the matrix size so as to meet the requirements of both computing efficiency and memory capacity.

Except for matrix A ($N \times N$), the operating system and other expenses will also occupy some memory. The communication in the meantime will use some cache. Based on previous experiences, 80% of memory occupation of HPL will get the optimal performance.

In practice, the size of N should be decided based on the actual software/hardware system. As for the ASC contest, the 3 kW power budget limits the system to a small scale with the number of nodes to be around 8–14, so choosing the size of N is less strict than on large-scale systems.

11.3.1.2 Size of Blocking

Decomposing the HPL matrix into blocks can improve the data locality and can be beneficial for the performance. Therefore, the size of block *NB* is an important parameter for the overall performance. The best value of *NB* may be different based on different software or hardware environments, and is usually determined through experimental measurement. However, there are some useful tips: *NB* should not be too large or too small, usually less than 512; *NB* is usually a multiple of 64.

11.3.1.3 Two-Dimensional Process Array

The two-dimensional process array ($P \times Q$) is determined based on the following rules:

- The value of $P \times Q$ should be equivalent to the total number of processes, which is assigned through using parameter $-n$ in command "*mpiexec −n 4./xhpl*".
- When only using the GPU cards in a heterogeneous architecture to compute HPL, the value of $P \times Q$ should equal the number of GPU cards, and the number of processes should also equal the number of GPU cards.
- $P \leq Q$ is based on the experimental experiences. In general, *P* should be set as small as possible, as the communication on the column direction is heavier than that on the row direction.

Note that the above suggestions are based on reference experience. Users should optimize those parameters based on the exact systems they are choosing.

11.3.2 Selecting the Third-Party Library

The third-party library mainly refers to MPI (such as MPICH2, IMPI, OPENMPI) and BLAS (such as ATLAS, GOTO, MKL). Almost all the HPL computation on CPU is done by the BLAS library, which means choosing the right BLAS library can be fundamentally important. To test HPL on the CPU + GPU platforms, it demands the GPU vendor to provide the BLAS library.

In the previous installations, we use MPICH2 and ATLAS libraries to finish the installation. Users can choose to use other different libraries based on their demand. As most teams of SCC choose Intel Xeon series CPU, MPI, and BLAS provided by Intel would be good choices to obtain ideal performance.

11.3.3 Compiler Optimizations

The compiler provides automatic optimizations to improve the performance without changing the programs. It is the easiest way to improve the performance. Therefore, users can choose to use compilers with a high optimizing capability and can configure between different compiler options.

As most teams of SCC use Intel Xeon series CPU, the C/C++ compiler icc/icpc, and the Fortran compiler ifort designed by Intel should be the best choices to obtain better performance. The non-commercial versions of those compilers can be downloaded from Intel® download center. Table 11.2 lists some options of icc.

Note that the compiler optimizations may have little effect on the HPL performance, as almost all the operations in HPL have been assigned to the BLAS library, and the compiler option usually cannot improve the performance of library. However, it is still an important optimization for other applications.

11.3.4 The Relation Between Node and Process

Section 11.3.2 mainly introduced test command within one node. If we want to test HPL based on multi-node clusters, it is necessary to point out the exact nodes to use, using the following command:

- *mpirun –np 3 -hosts node01 node02 node03*

By default, MPI sequentially uses the nodes input in the command to allocate processes. Therefore, process 0, 1, and 2 will be assigned to node 01, node 02, and node 03, respectively based on the command above. Changing the sequence of input nodes may sometimes greatly effect the performance and the stability of the system.

Table 11.2 Some options of icc

Option	Function
–O2	Maximize the execution speed code. Including the global instruction scheduling, software pipeline, prediction, pre-fetch, etc.
–O3	Based on –O2, add more aggressive loop and memory access optimizations, such as scalar replacement, loop unrolling, code replication, loop blocking, etc.
–ax_	Appoint the generated code to special processors, generate the IA-32 instruct at the same time
–funroll-loops	Loop unrooling
–static	link static library
– fomit-frame-pointer	Turn on EBP as general register

In experience, the following aspects should be noted:

(1) Do not use the shared file node as process 0, as the node takes charge of the shared file management;
(2) When the performance is worse than expected, changing the sequence of two poorly effective nodes might be helpful;
(3) The task of HPL may not be allocated equally. Therefore, the task assignment should be based on the size of the task.

11.3.5 Task Allocation Optimizations

Task allocation on CPU + GPU heterogeneous clusters is more complicated than that on CPU-only homogeneous node. On heterogeneous platforms, CPU and GPU work simultaneously to finish DGEMM. In other words, part of the matrix multiplication is done in the GPU by using the DGEMM function from the NVIDIA CUBLAS library; while at the same time, the other part is done in CPU by using the DGEMM function from the BLAS library. Therefore, it is necessary to decide the task allocation between CPU and GPU.

11.3.5.1 DGEMM Parallelization on CPU and GPU

Figure 11.2 introduces two decomposing methods (column decomposition and row decomposition) to complete the multiplication of matrix A $(M \times N)$ and matrix B $(N \times K)$ and obtain matrix C.

The two decompositions are based on two situations of the parameters M, N, K: the first is that when K is a small value and M, N are large values, shown on the left of Fig. 11.2; and the second is that when K and N are equal and small values, while M is a large value, as shown on the right of Fig. 11.2.

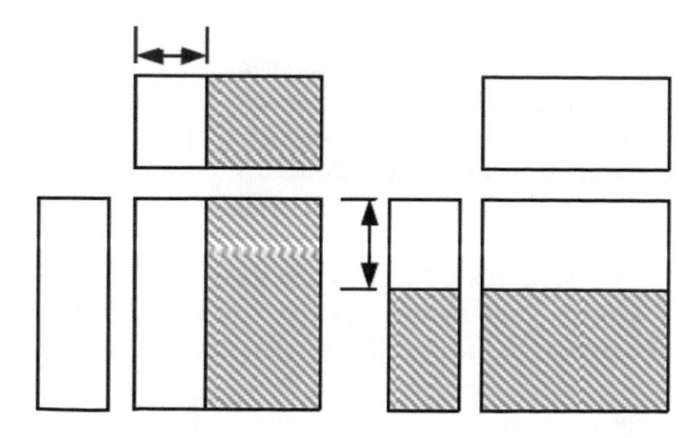

Fig. 11.2 Left: column decomposition, Right: Row decomposition

Adjusting the ratio R can achieve the load balance between CPU and GPU. The best situation is that the computing time of CPU equals the computing time of GPU plus the data transfer between CPU and GPU.

There are two points to notice about the implementation details:

(1) The DGEMM function in CUBLAS uses the algorithm by Volkov et al., which is sensitive to the value of M, N, and K. In general, the best performance is achieved when M is the multiple of 64, and N and K are multiples of 16.
(2) GPU may not be effective in small-scale tests. Therefore, users can directly use the CPU DGEMM when the scale does not surpass the threshold.

11.3.5.2 Load Balance Between CPU and GPU

In (1), the GPU and CPU use a static task decomposition mechanism to finish the DGEMM and DTRSM, and the ratio R is fixed. While HPL itself is an iterative process, the computing scale will decrease with the iteration going forward. The performances of CPU and GPU will be decreasing with the size reduction of the matrix. Therefore, it is necessary to use a dynamic task decomposition mechanism that can update the value of R based on the real performance situation.

Note that dynamic task decomposition mechanism may require code modification.

11.3.6 Heat Dissipation

Besides the optimizations introduced above, there are other factors that deserve our attention. For example, the work condition of GPU is very sensitive to the temperature. Therefore, we need to pay attention to the heat dissipation of the system. We have noticed that once the temperature of GPU went too high, the performance would decrease greatly, or the system would even break down.

The first solution is to strengthen the dissipation by setting the BIOS to adjust the fan speed. The second solution is to aggressively change the fan channel. There are real cases in which the fan channel was changed so that the wind went more directly to the GPU cards and cooled them down. By doing this, they managed to increase the performance to 40 GFlops per card, and greatly decreased the possibility of system breaking down.

During the competition, factors like the weather and site can cause a problem of dissipation, and lead to lower performance than expected. For example, in the SC13 in November 2013, 2 teams found their performances were not stable, the reason of which was later discovered to be the fan's backing to the cards. After rotating the direction of the fan by 90°, the performance became stable.

Nevertheless, support from the device vendors to provide platforms with more effective cooling systems may be the fundamental way to solve the problem of dissipation.

11.3.7 Further Reading

- **ATLAS** can be downloaded from http://sourceforge.net/proposals/math-atlas/files.
- **HPL** source code can be downloaded from http://www.netlib.org/benchmark/hpl/.

Chapter 12
Optimization for the Molecular Dynamics Software GROMACS

In this chapter, we will first give a brief introduction about the GROMACS scientific computing software, and its experiment results based on a four-node HPC cluster. Furthermore, we propose and analyze the optimizations for GROMACS.

12.1 Introduction

GROMACS (GROningen MAchine for Chemical Simulation) is developed based on classical molecular dynamics theory. GROMACS is able to simulate the particle collections of solid, gas and liquid as well as the dynamic reaction of molecular under any force fields or any boundary conditions. The simulation scale varies between hundreds and millions of molecular. GROMACS is widely applied in biochemical molecular studies such as the studies of proteins, lipids, nucleic acids and so on.

GROMACS is implemented by using MPI (message passing interface) to parallelize the molecular dynamic software for parallel computing systems. Bounded forces, unbounded forces and external forces are the three kinds of forces that can be simulated by GROMACS. Molecular dynamic simulation is achieved by using leap-frog and Verlet for Newton Coupling Equations and using other numerical methods for boundary conditions. Domain decomposition is applied to allocate the computing task into different compute nodes. However, the computing efficiency is usually largely affected by the load imbalance due to the difference of communication granularity and the asymmetry molecular distribution.

To achieve a better performance, GROMACS makes some optimizations on both the equations and the algorithm. For example, GROMACS has its own library to compute the reciprocal after the square root operation, and its intrinsic function can make the compiler achieve better parallelism from the instruction level. Since version 4.6, GROMACS have provided support for CUDA to achieve GPU acceleration.

ASC Community, *The Student Supercomputer Challenge Guide*,
https://doi.org/10.1007/978-981-10-3731-3_12

GROMACS is a free software under the GNU standard. In the following experiment, we will use the version 4.6.1 to demonstrate the performance test of the "DPPC in water" case officially provided by the GROMACS.

12.2 Experimental Environment and Hardware Configuration

The HPC computer we choose has four compute nodes connected by InfiniBand interconnection. Each node has two 10-core Intel CPUs, and an NVIDIA Tesla K20X accelerator. Table 12.1 shows the basic parameters.

12.3 Parameters

In this section, we analyze the influences of different parameters on performance. Only a small number of the input parameters can affect performance, such as the 'nsteps' that refers to the total steps of the simulation. As the time for each step is fixed, the overall execution time is proportional to nsteps. Thus in Fig. 12.1, the overall time is 14 s when nsteps = 1000, while it is 64 s when nsteps increases to 5000. Therefore, the execution time of GROMACS can be estimated as follows:

Execution time = $\alpha \times$ nsteps + β (where $\alpha = 0.0127$, $\beta = 1.32$ in Fig. 12.1).

The following parameters in GROMACS can also affect the execution time as they are used to control the step interval of output:

- nstxout: the step interval for outputting the coordinate.
- nstvout: the step interval for outputting the velocity.
- nstxtcout: the step interval for outputting the position.
- nstlog: the step interval for outputting the energy.

Table 12.1 Basic parameters of one node

Item	Name	Configuration	Number
Server	CPU	Intel® Xeon® CPU E5-2670 v2 @ 2.50 GHz, 10 core	2
	Memory	16G, DDR3, 1600 MHz	10
	Hard disk	300 GB, SAS, 15,000 rpm	2
Accelerator card	GPU	NVIDIA Tesla K20X (2688 cores, 6 GB GDDR5 Memory)	2
HCA card		InfiniBand Mellanox Technologies MT27500 Family [ConnectX-3]	1

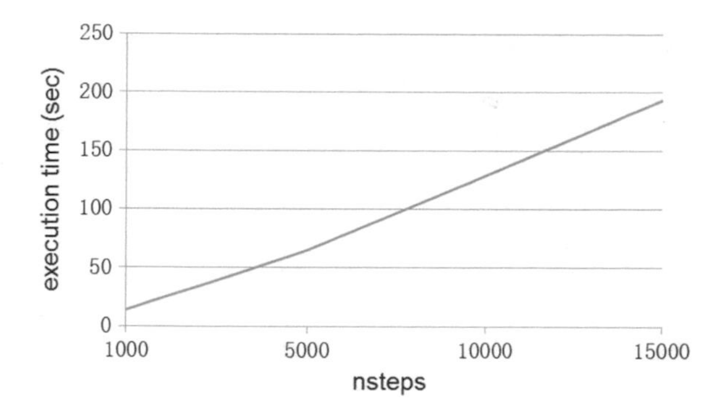

Fig. 12.1 Relation between execution time and nsteps

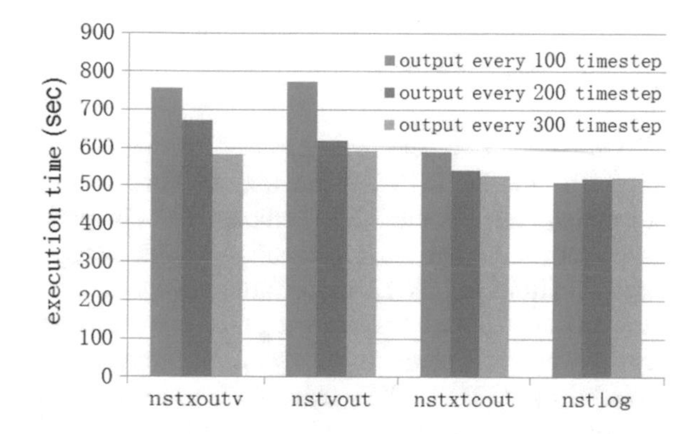

Fig. 12.2 Influence of outputting parameters on the performance

Figure 12.2 demonstrates the execution times based on different step intervals of output step. The value 100 means that there is an output every 100 time steps. Therefore, decreasing the step interval means increasing the output frequency. Figure 12.2 shows that increasing the step interval can decrease the execution time, as outputting the results demands extra cost. Practically, it is better to set it to zero if outputting the information is not necessary.

12.4 Performance Analysis of the Software Library

GROMACS needs to use many third-party software and libraries, including the compiler, MPI, and scientific computing libraries such as Fast Fourier Transform. In this section we will introduce the influences of using different software and libraries on the execution time. Here we set the nsteps to be a fixed number, and turn off all the output related parameters.

12.4.1 Compiler

The compiler can affect the overall performance. As we are using the Intel CPUs, we can choose compilers developed by Intel, apart from using the GNU compiler. The comparison of the two compilers is shown in Fig. 12.3, in which we can find out that Intel compiler can decrease the execution time by around 20% over the GNU compiler based on either 8-core or 16-core configurations. When using 8 cores, the execution time for GNU and Intel compilers is 6710 s and 5409 s, respectively. When using 16 cores, the execution time is 3827 s and 3102 s. The Intel compiler generally has 20% higher efficiency. Therefore, it is useful to choose an Intel compiler for Intel machines.

12.4.2 MPI Tool

GROMACS uses MPI for parallel computing among nodes. OpenMPI and Mvapich2 are the MPI implementation libraries that are widely used in Top500 supercomputers. Figure 12.4a demonstrates comparisons of different MPI libraries based on 8-core and 16-core configurations. Mvapich2 can slightly decrease the execution time. In Fig. 12.4b we further show the performance on different numbers of 8-core nodes. With the increase of the compute nodes, the cost on communication also increases. Therefore the speedup increases from 6% (Mvapich2 = 3102, OpenMPI = 3321) to 20% (Mvapich2 = 1274, OpenMPI = 1596). Obviously for Mvapich2, using more nodes can result in a more efficient network.

12.4.3 Fast Fourier Transform (FFT) Library

FFT is frequently used in GROMACS. We choose two libraries, FFTW and MKL to compare their performance differences. As shown in Fig. 12.5, MKL is only slightly slower than FFTW on both the 8-core and 16-core CPU. Thus the FFT library has little influence on the GROMACS performance.

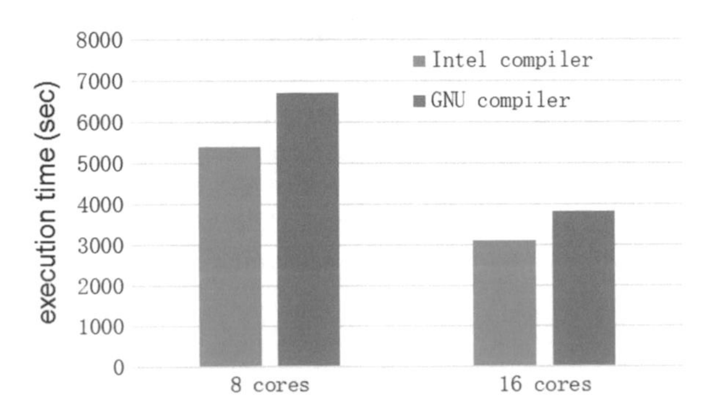

Fig. 12.3 Comparisons of different compilers

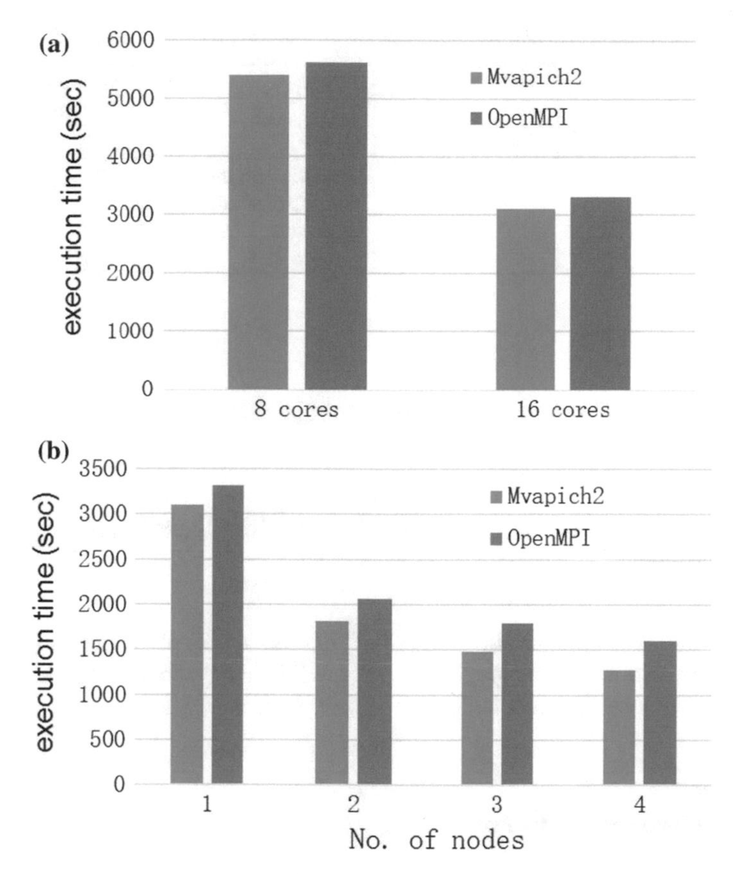

Fig. 12.4 **a** Comparison of different MPI versions on different number of nodes. **b** Comparison using different MPI on different number of nodes

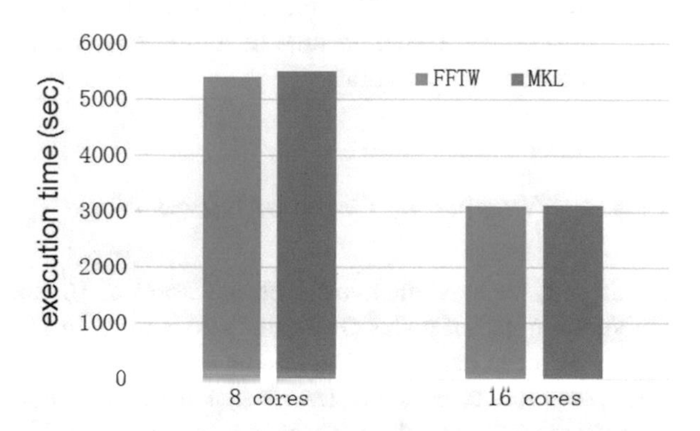

Fig. 12.5 Performance comparison of FFT library on different cores

Table 12.2 Summary of the best configuration

	Compiler	MPI library	Fast fourier transform library
Optimal configuration	Intel compiler	Mvapich2	FFTW
Speedup	20%	Improve with the increasing of cores 1node 6% 4 nodes 20%	<1%

12.4.4 *Summary*

Table 12.2 gives the summary of the best configuration:

12.5 Performance Analysis of Computing Resource Allocation

Based on the best configuration described earlier, we explore the influences of adjusting the resource allocation. Three parts will be analyzed: (1) number of cores, (2) number of cores on each node, and (3) number of nodes.

12.5.1 *Number of Cores*

We first tested the performance based on a single node. The result is shown in Fig. 12.6a. With the number of cores increasing from 2 to 16, the execution time is decreasing accordingly. The speedup factor is shown in Fig. 12.6b, which is the ratio between the execution time of a dual-core over the execution time of other corresponding configurations.

12.5.2 *Number of Cores on Each Node*

In this part, we fixed the number of total cores at 16, and test the performance using different number of nodes (1, 2 and 4). As shown in Fig. 12.7, the execution time is increased when using more nodes, as more inter-node communication is required. The execution time is the least when using a single node because there is no inter-node communication. Therefore, parallelism within a single node is generally better than parallelism over more nodes. It is practical to employ all the cores inside one node for GROMACS so as to make full use of the computing resources, and decrease inter-node communication.

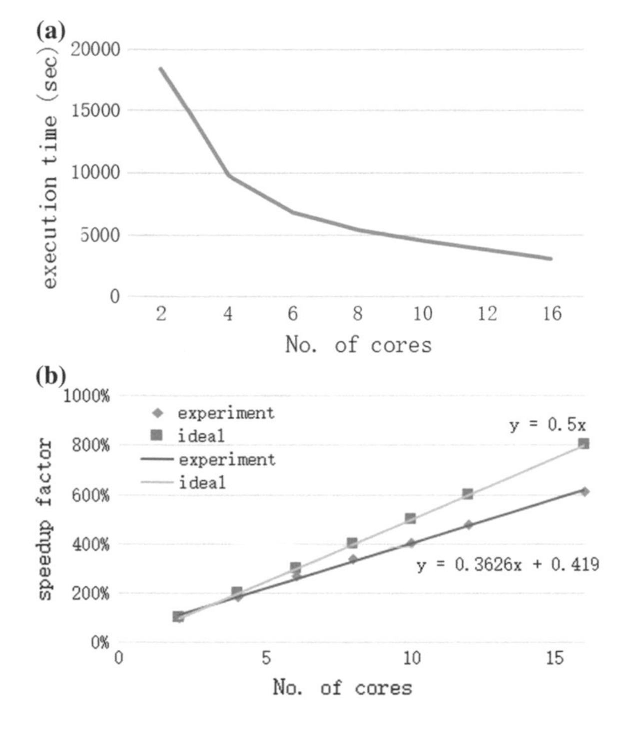

Fig. 12.6 **a** Execution time on single node. **b** Speedup on single node

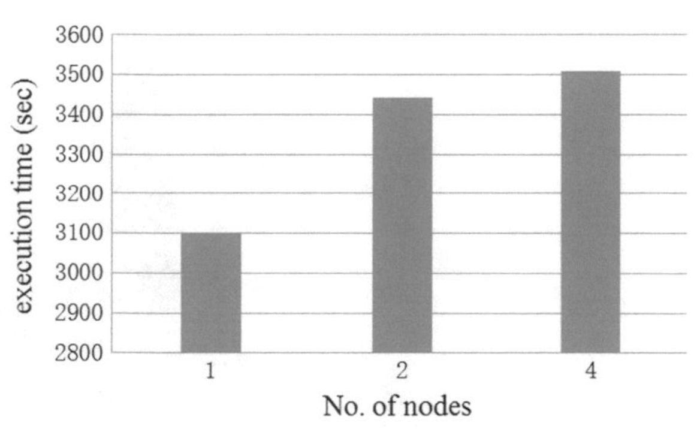

Fig. 12.7 Execution time using 1, 2 and 4 nodes (total number of cores is fixed)

12.5.3 Number of Compute Nodes

In this part, we fixed the number of cores inside each compute node, and tested the performance over different numbers of compute nodes. As shown in Fig. 12.8a, with an increase in the node number, the execution time is decreased. However, the parallel efficiency is decreased with the increase of node number, as shown in Fig. 12.8c. The parallel efficiency is 60% when using four nodes. It once again proves that increasing the number of compute nodes will generate inter-node communication, which is very important to the performance.

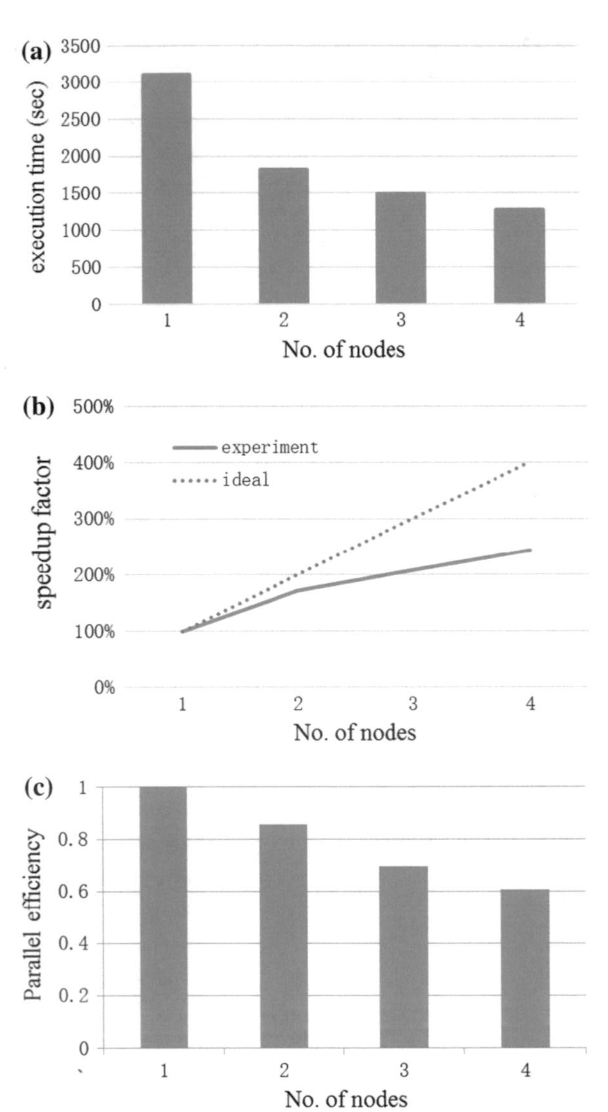

Fig. 12.8 **a** Execution time on different number of nodes. **b** Speedup on different number of nodes. **c** Parallel efficiency on different number of nodes

12.5.4 Summary

The performances above show the influences of different numbers of resources and different configurations. As GROMACS will have extra communication between different processes, the network bandwidth is important to determine the performance, as well as the full exploration of single-node cores. In general, GROMACS has good performance scalability, and the parallel efficiency is fixed. Therefore, we can decrease the execution time by increasing the number of both the cores and nodes.

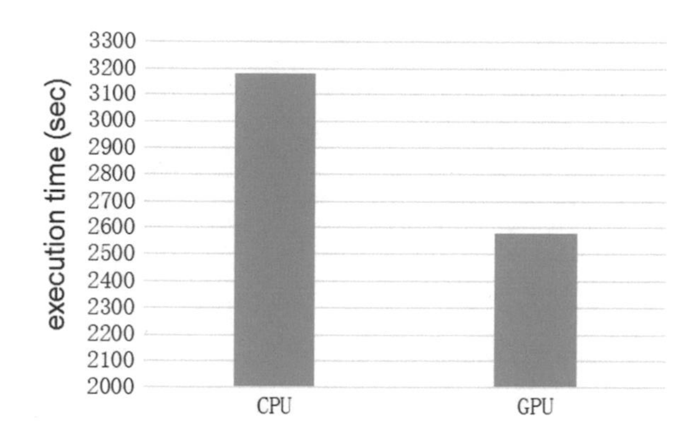

Fig. 12.9 Performance comparison between CPU and GPU

12.6 Performance Test Using GPU

Since version 4.6, GROMACS provides support for GPU CUDA. One CPU core is related to one GPU card using OpenMP thread to communicate. In our experiment, as we only have one GPU inside one node, only one CPU core will be involved for computing. Version 4.6 is the first to support CUDA, so that there is a lot to improve, and we can expect better support for later versions. Figure 12.9 shows the performance comparison between using GPU and using 16 CPU cores. The performance using one GPU core is better than using 16 CPU cores, which means GPU is very suitable to use. However, the performance becomes worse when using more than one GPU, which is caused by the bandwidth bottleneck from the InfiniBand card connecting the GPU with other nodes. NVIDIA has proposed to decrease the I/O path in its new hardware design and drive approaches. There is also related work on this part. Therefore, the communication between GPU cards would become the key factor determining overall performance. Even though such a bottleneck cannot be eliminated, the performance improvement of GPU is also very promising over CPU, and thus it is necessary to use GPU in future accelerations.

12.7 Further Reading

- **GROMACS** The software and instruction can be accessed from http://www.gromacs.org/.

Chapter 13
Optimizations for Ocean Model LICOM

13.1 Introduction to LICOM

13.1.1 LICOM

LICOM[14-1] (LASG/IAP Climate system Ocean Model) is an ocean model in the climate system developed by the State Key Laboratory of Numerical Modeling for Atmospheric Sciences and Geophysical Fluid Dynamics (LASG), Institute of Atmospheric Physics (IAP), and the Chinese Academy of Science. LICOM is designed by the global ocean-atmosphere coupler group to simulate large-scale wind circulation and thermohaline circulation. Besides the ocean model, there are other models such as the atmosphere model, ice mode, and land model. They are connected by the coupler to form a complete climate model. LICOM is a big contribution by LASG to the development of the climate system.

From the view of the computer, LICOM is a program used to solve the numerical solution of partial differential equations. The program simulates the dynamic circulations of the ocean wave, thermal, and salt. LICOM is written in Fortran90 and can be scaled over different kinds of parallel computers or computing clusters.

LICOM has 59 F90 source files, and 1 C source file, with 18,000 lines of code. Licom.F90 is the main program that executes other subprograms.

13.1.2 The Development of LICOM

In the past decade, LASG has released ocean models with 4 level, 20-level, and 30-level versions. LICOM is updated very frequently. The first version is called LICOM 1.0; it is a horizon lattice 0.5° × 0.5°, vertically 30 level, and covers the domain from 65°N to 75°S. It is derived from the original version with a horizon

ASC Community, *The Student Supercomputer Challenge Guide*,
https://doi.org/10.1007/978-981-10-3731-3_13

lattice 1.875° × 1.875° and is vertically 30 level. The version used in the 2014 ASC competition is the latest LICOM 2.0.

13.1.3 Necessity of LICOM Acceleration

Nowadays, even though the living standard is greatly improved by technology advancement, human activities are still largely influenced by the climate. Especially in the twenty-first century, global climate change has brought significant challenges to human life. Although climate change is related to the whole climate system, the dynamic circulation of the ocean is essential in the scale from year, decades, to centuries. Therefore, as the key tool to help us study the ocean, ocean models such as LICOM are necessarily important. As for the development, study, and the research of LICOM, there is still much to do to parallelize the program, and such work is essential.

13.2 Analysis of the Key Optimization Points of LICOM

LICOM optimization is one of the topics in ASC14, and aims to test the performance of the cluster and the software environment built by each team. Besides, the parallelization strategy is the essential factor. We list some key points for LICOM optimization:

- Achieving the maximum performance within the 3000 W power budget.
- Using compilers that are more suitable for Intel Xeon platform, such as the Intel compilers, Intel MPI;
- Learning Fortran90 programming language, the main program licom.F90 and the subprograms, and the shell script.

13.3 Hardware Environment

Combining all factors in the LINPACK test and other topics including LICOM, we choose to use the hardware platform shown in Table 13.1, to maximize the performance.

Table 13.1 The hardware environment of Taiyuan University of Technology (TUT) during ASC14

	Name		Number
Server	Inspur NF5280M3	CPU: Intel Xeon E5-2692v2 × 2, 2.2 GHz, 12 cores Memory: 16G × 8, DDR3, 1600 MHz disk: 1 TB SATA × 1	10
HCA card	FDR		10
Switch	FDR-IB		1
Cable	IB		10

13.4 Software Environment Configuration

Based on the hardware platform mainly from Intel and previous optimizing experience the final software configuration is shown in Table 13.2:

The next important step is to install NetCDF[14-2] and the LICOM program.

13.4.1 Install and Configure the NetCDF C Library

Create the LICOM directory (~/LICOMtest/ASC14_LICOM/) and the NetCDF install directory (~/local/netCDF/), the latest NetCDF C (netcdf-4.3.1.1) is used, install based on the following steps:

(1) Configuration (note there are many options besides the one we use here):

```
[Speedyfeng@node1 netcdf-4.3.1.1]$ ./configure –prefix = /home/Speedyfeng/local/netCDF –disable-netcdf-4 –disable-dap
```

(2) Make:

```
[Speedyfeng@node1 netcdf-4.3.1.1]$ make check install
```

Table 13.2 The software configuration of TUT during ASC14

	Name	Version
OS	Red Hat Enterprise Linux Server release 6.2 (Santiago)	2.6.32-220.el6.x86_64
Compiler	Intel compiler	12.10
	GUN compiler	4.4.4
Database	Intel MKL	11.1.0.080
MPI	Intel MPI	4.1.3.049
	MPICH2	2-1.5

(3) Edit the .bashrc file to add the /home/Speedyfeng/local/netCDF/lib into the library path:

```
export LD_LIBRARY_PATH = $LD_LIBRARY_PATH: /home/Speedy
feng/local/netCDF/lib
```

13.4.2 Install the NetCDF Fortran library

Based on the NetCDF C, we use the latest netcdf-fortran-4.2 to configure NetCDF Fortran library as follows:

(1) Configure:

```
[Speedyfeng@node1 netcdf-fortran-4.2]$ CPPFLAGS = -I/home/
Speedyfeng/local/netCDF/include LDFLAGS = -L/home/Speedy
feng/local/netCDF/lib ./configure –prefix = /home/Speedy
feng/local/netCDF
```

(2) Make:

```
[Speedyfeng@node1 netcdf-fortran-4.2]$ make check
install
```

(3) Set the NetCDF environment variables:

```
export PATH=$PATH: /home/Speedyfeng/local/netCDF/bin
export NETCDF=//home/Speedyfeng/local/netCDF
```

13.4.3 LICOM Installation

After correctly installing the software environment such as the Intel compiler, Intel MPI, and the NetCDF, we edit the case.sh file of LICOM 2.0. We need to edit the main directory, the run mode, the number of processes, the compiler, the parallel library, the NetCDF library, and so forth (we don't give an extra introduction to the OpenMP of LICOM 2.0). The case.sh after modification is as follows:

```
#set TESTNUM=01
setenv CASENAME test01  # based on the steps of test, set to 01、02、03…
setenv LICOMROOT /home/Speedyfeng/LICOMtest/ASC14_LICOM # LICOM dir
setenv SRCPATH  $LICOMROOT/src
setenv BLDPATH  $LICOMROOT/bld
setenv DATAPATH $LICOMROOT/data
setenv EXEROOT  $LICOMROOT/$CASENAME
setenv EXESRC   $EXEROOT/src
setenv EXEDIR   $EXEROOT/exe
setenv RUNTYPE  initial                # run type:continue or initial
set HISTOUT = 1                       # model historic output, no use for this
version !!!!!
set RESTOUT = 1                        # model restar file outpur every day
set XNTASKS = 10
set YNTASKS = 20
set NTHRDS  = 1
set LID = "`date +%y%m%d-%H%M%S`"
@ NTASKS = $XNTASKS * $YNTASKS

echo '-------------------------------------------------------------'
echo '    Produce Makefile                         '
echo '-------------------------------------------------------------'

\cat >! Makefile << EOF
################################################################
# modify this part according to compiler and system you are using
# CPPFLAGS CPP INCLUDE NLIB FFLAGS FC
################################################################
CPPFLAGS = -P -traditional
CPP     = /usr/bin/cpp \$(CPPFLAGS) \$(INCLDIR)
INCLDIR          =      -I.      -I/usr/include      -I/usr/local/include
-I/home/Speedyfeng/intel/impi/4.1.2.040/include
-I/home/Speedyfeng/local/netCDF/include
NLIB    =          -L/home/Speedyfeng/intel/impi/4.1.2.040/lib64    -lmpi
-L/home/Speedyfeng/local/netCDF/lib -lnetcdf -lnetcdff
FFLAGS =  -mcmodel=large -shared-intel -O2 -r8 -i4  -convert big_endian
-assume byterecl -no-vec
FC = /home/Speedyfeng/intel/impi/4.1.2.040/intel64/bin/mpiifort \$(FFLAGS)
################################################################
```

```
\cat >! Makefile << EOF
######################################################################
# modify this part according to compiler and system you are using
# CPPFLAGS CPP INCLUDE NLIB FFLAGS FC
######################################################################
CPPFLAGS = -P -traditional
CPP      = /usr/bin/cpp \$(CPPFLAGS) \$(INCLDIR)
INCLDIR  = -I. -I/usr/include -I/usr/local/include -I/home/Speedyfeng/intel/impi/4.1.2.
040/include -I/home/Speedyfeng/local/netCDF/include
NLIB =   -L/home/Speedyfeng/intel/impi/4.1.2.040/lib64 -lmpi -L/home/Speedyfeng/local/n
etCDF/lib -lnetcdf -lnetcdff
FFLAGS = -mcmodel=large -shared-intel -O2 -r8 -i4 -convert big_endian -assume byterec
l -no-vec
FC = /home/Speedyfeng/intel/impi/4.1.2.040/intel64/bin/mpiifort \$(FFLAGS)
######################################################################
OBJS   = precision_mod.o param_mod.o pconst_mod.o pmix_mod.o isopyc_mod.o cdf_mod.o\
tracer_mod.o dyn_mod.o work_mod.o output_mod.o forc_mod.o msg_mod.o sw_mod.o canuto_mod
.o\
```

Fig. 13.1 The key configuration part of case.sh

The key configuration part of case.sh is shown as Fig. 13.1.

After the modification of the case.sh using vim, we can use "./case.sh" to run the script to do the compilation, link operations, and generate the test directory. The licom2 file has been generated in test/exe/directory and now we can test the computation for different loads.

13.5 Implementation and Result

13.5.1 The Implementation

As for LICOM 2.0, all teams use the MPI to compute in the ASC final. Based on the computing platform with 10 24-core nodes, the final configuration is mpiexec –ppn 20 –n 200 ./licom2<ocn.parm. The mpdring in node1 to node 10 has been previously mpdbooted, which proves good performance for load A, B, and C, and the power sustained around 1960 W. However, for load D, the power soon exceeds 3000 W, so that we instantly change to the backup method that sets the total number of processes to 180. The MPI strategy is mainly based on the rules of licom. F90's approach of calling other subprograms, as well as the practical experience after running the program many times.

```
module param_mod
#include <def-undef.h>
use precision_mod
integer,parameter:: jmt_global=JMT_GLOBAL  ! Number of the End Grid for Tracer
in Latitude.
integer,parameter:: jmm_global=jmt_global-1
integer,parameter:: jstart    = 3     ! Satrting grid for Tracer in Latitude.
integer,parameter:: imt_global=362   ! Number of Grid Points in Longitude
integer,parameter:: km=30     ! Number of Grid Points in Vertical Direction
#ifdef SPMD
integer,parameter:: nx_proc=NX_PROC ! Number of MPI tasks in zonal direction
integer,parameter:: ny_proc=NY_PROC ! Number of MPI tasks in meridional
direction
integer,parameter:: n_proc=nx_proc*ny_proc ! Total number of Processors for
MPI
integer,parameter:: num_overlap=2 ! Number of overlapping grids for
subdomain.
integer,parameter:: jst_global=jstart ! Number of the Strating Grid for Tracer
in Latitude.
!Nummber of grids in the each subdomain
integer,parameter:: jst=1     ! Number of the Strating Grid for Tracer in
Latitude.
integer,parameter:: jsm=jst+1 ! Number of the Strating Grid for Momentum in
Latitude.
integer,parameter::
jet=(jmt_global-jst_global+1-num_overlap)/ny_proc+1+num_overlap
integer,parameter:: jem=jet-1 ! Number of the End Grid for Momentum in
Latitude.
integer,parameter:: jmt=jet
integer,parameter:: imt=(imt_global-2)/nx_proc+num_overlap
integer :: j_loop           ! Loop index of J cycle for the each subdomain.
#else
integer,parameter:: jst=jstart ! Number of the Strating Grid for Tracer in
Latitude.
integer,parameter:: jsm=jst+1 ! Number of the Strating Grid for Momentum in
Latitude.
integer,parameter:: jet=jmt_global  ! Number of the End Grid for Tracer in
Latitude.
integer,parameter:: jem=jet-1 ! Number of the End Grid for Momentum in
Latitude.
```

```
integer,parameter:: jmt=jmt_global
integer,parameter:: imt=imt_global
integer,parameter:: nx_proc=1
integer,parameter:: ny_proc=1
#endif
```

The definition of nx_proc=NX_PROC (number of MPI tasks in zonal direction) and ny_proc=NY_PROC (number of MPI tasks in meridional direction), is similar to the definition in case.sh (#define NX_PROC $XNTASKS, #define NY_PROC $YNTASKS). Therefore, setting XNTASKS and YNTASKS to be divided by 360 is good for decomposition (if the values of XNTASKS and YNTASKS are not set correctly, it may cause error during execution). Compared with similar hardware platform and environment configuration, the bigger the product of XNTASKS and YNTASKS (i.e., NTASKS, number of MPI processes) is, the faster the execution speed.

As there are 2 12-core CPUs in each of the 10 compute nodes, the best performance is achieved when XNTASKS=10, YNTASKS=20. Detailed information is given in Sect. 13.7.

13.5.2 Optimization Results

After fully considering the discussions above and the proposal, we finish the optimization of LICOM on four loads before noon. Specific information is shown as follows:

(a) Load A uses the initial mode. We copy the generated licom2 executable to the directory of load A. 200 MPI processes are used and the power is close to 3000 W. The execution time is 1008.17 s;
(b) Load B copy file fort.22.0002-01-01 from load A and uses the continue mode. We copy the generated licom2 executable to the directory of load B. 200 MPI processes are used and the power is close to 3000 W. The execution time is 534.85 s;
(c) Load C uses the initial mode, and we copy the generated licom2 executable to the directory of load C. 200 MPI processes are used and the power is close to 3000 W. The execution time is 1437.98 s;
(d) Load D also uses the initial mode, but the operation is more complicated. Based on the requirement in README, we copy the recompiled licom2 executable to the directory of load D. 180 MPI processes are used, and the power is close to 3000 W. The execution time is 3794 s.

13.6 Review of Our Preparation for ASC14

I was invited to be the advisor of the participating team in the ASC14 by Professor Ming Li, the dean of the School of Mathematics in Taiyuan University of Technology (TUT), in late 2013. At that time, the team was not yet officially formed. I felt it would be a big challenge but also a big opportunity for me, and also for the students to join the team.

2014 ASC contained a preliminary stage (from January 2, 2014 to February 27, 2014), and the final stage (April 20, 2014 to April 25, 2014). During the preliminary stage, each team submitted their proposal to the committee, who would later determine the teams for the final stage. The teams entering the final stage used the supercomputers provided by Inspur and optimized the programs according to the tasks given by the committee.

The first difficulty for me was to find the best team members who were interested in this new HPC area. Professor Li suggested that I should not only form the competing team for this year, but also the echelon for future competitions. Therefore, with the help of the student union, we published a notice on the website and recruited a group of students with interest, enthusiasm, and ability. They are from different schools, majors, and grades, but with one thing in common, the passion for supercomputing.

Based on the requirements of the competition, I suggested the team made a long-term study plan. I told them first to do some preparations so as to fully consider and identify what was the difficulty and where to find the breakthroughs. In other words, it is very important to know the rules, the procedures, to find the knowledge we need to learn as well as to analyze the hardware and software environment. We needed to contact the organizing committee whenever there was something we were not sure about. All those preparations required us to make specific plans, as well as the implementation of the plans. The team leader, Xichen Cao was in charge of the weekly meeting to summarize and share the experience among the group.

To be better prepared, each team member was assigned different topics, so that each one could do his best in the team. There is a common room in the TUT cloud computing center that had a table, three sofas, and a whiteboard. This common room became the regular place for the team to meet and discuss, named the "temple of thoughts" by the team. They also created a QQ group to facilitate the sharing of materials. Everyone in the QQ group had the authority and obligation to upload and download material. They used "guess computing," "hand computing," "abacus computing," "machine computing," and "supercomputing" to label different QQ IDs, which was quite creative and corresponded to the theme of the competition. The preparation was full of passion, and everyone in the team enjoyed their time.

During the winter vacation between the preliminary stage and the final stage, the team members had a good rest, but the team spirit might also have become slack. To keep the pace and passion, we introduced the knocking-out mechanism. We would determine the students to participate in the competition based on their performance. I think it not only provided a fair opportunity for all students enrolled, but also guaranteed the quality of the final team. Most of the students only took a one-week vacation, and came back to the university to continue the preparation.

The preliminary stage required us to provide the proposal for the design of the cluster. Even though we could access the remote platform provided by the organizing committee, the students still needed to get familiar with the set up process of real systems. With the help of Prof. Ming Li, we borrowed two compute nodes from the TUT cloud computing center, so that the students could practice in a real environment.

3D_EW accounts for the biggest part in the total points. We only made a little progress and could not use MIC due to the size of the data sets. During one discussion, one student Yanjun Gao used a cube to describe the data set, and it inspired the captain Chenxi to come up with a great idea to reduce the data-set size. Based on this idea and the later optimization using SIMD and MPI dynamic scheduling, the team kept working day and night for a few days. At last, the team managed to achieve 400× speed up which was a first among mainland universities. Finally we were selected for the final round.

We were quite excited to be able to enter the final stage, but were also anticipating the intense competition to come. I told the students to participate with the most positive attitude, and to participate with the purpose of learning from others. The final rank would not matter. The most important thing was that we got this opportunity to compete with other experienced teams from the top universities from around the world.

13.7 Review of Optimizing LICOM in ASC14 Final

As the dark horse in the final stage, we did our best to achieve good results. Jiangfeng Zhang in our team was in charge of the LICOM optimization.

On April 21, the first day of debugging before the formal competition, the work we did included removing the MIC card, building the cluster, and installing the Linux system, IB network, and NFS. In the afternoon, we encountered a problem that the mpd ring could not be successfully created, which meant that the MPI was unable to be used. When other teams left in the evening, Jiangfeng Zhang insisted on staying for the problem. We stayed late that night and finally found the solution to the problem.

On April 22, we were excited to find that the MPI was finally running successfully. We instantly started to test the LICOM application. After testing different configuration (XNTASKS = 10, YNTASKS = 24, XNTASKS = 8, YNTASKS = 24, XNTASKS = 10, YNTASKS = 20, XNTASKS = 20, YNTASKS = 10, XNTASKS = 10, YNTASKS = 18, XNTASKS = 18, YNTASKS = 10, XNTASKS = 9, YNTASKS = 20, XNTASKS = 20, YNTASKS = 9), we finally used XNTASKS = 10, YNTASKS = 20 as the first choice, and XNTASKS = 10, YNTASKS = 18 as the backup choice. We managed to save the debugging time by summarizing the number and speed of computing the integral files, and predicting the remaining execution time based on the executed time shown in top command. Jianfeng Zhang wrote a file called “operation plan for tomorrow” one day earlier, which was helpful in saving time in the LICOM optimization. His file contained a detailed explanation for the setup of the “case.sh”, analysis on the MPI parallel strategy, wrote a script LICOM-A.sh to run the “mpiexec –ppn 20 –n 200 ./licom2 <ocn.parm”, and finished the time record command Speedyfeng@node1 A]$/usr/bin/time -p./LICOM-A.sh>TUT.LICOM.A.TIME-200-001.TXT 2>&1.

April 23 was the first day of the formal competition. The tests included LINPACK, QE, SU2, and 3D_EW. As we were an obvious distance behind the other universities in the LINPACK test, we decided to run LINPACK as fast as possible so as to have more time for other tests. We managed to persuade Yunxi Zhang to do so, even though he had focused on optimizing LINPACK for half a year. By running LINPACK for only minutes, we saved a lot of time for other tests. 3D_EW was in the charge of Chenxi Zhang, and we were extremely competitive. Due to the effort in both the preliminary and final rounds, we achieved a good performance in 3D_EW.

April 24 was the second day of the final competition. The tests included the LICOM and the mystery test. Everything went well with LICOM except for a small problem. Maybe because he was too nervous, Jianfeng Zhang missed some requirements of README and still set the mode to be initial for load B. Based on the error information, we set it back to continue but were informed that the fort.22.0002-01-01 file could not be found. We thought that the problem might be from the topic itself so we asked the organizing committee. It turned out to be our mistake. The story told us that there were pros and cons if you focused too much on one thing.

April 25 was the presentation day. Our draw was good. Based on our own situation, in our presentation, we focused more on the 3D_EW and LICOM, and the presentation was really successful. After that, everyone was relieved. In the afternoon, there was no doubt about our excitement when we were standing on stage to be awarded the cup.

During the days and nights in Guangzhou, I was so deeply impressed by the efforts and unity of our team. The time was so tight so we took turns to eat. We were always the last team to leave to grasp every second. We stayed late at night and got up early in the morning. Chenxi even stayed up for a whole night to debug before running 3D_EW in Tianhe-2, and we finally got the second place in Tianhe-2. Everyone stayed late the last night for preparing the presentation. Youlin Tian, who was in charge of giving the presentation practiced many times. The efforts of the students formed a good foundation for our success.

In conclusion, we need to thank the support from our university, the collaboration of the team, and the tremendous efforts of every team member. Setting the right task and role for each team member was extremely important. No one member could solve all the problems alone. As a fresh team, we had to compete by collaboration. It was a great experience for all the team members. Before the competition, the team members not only worked together, but also organized fun activities together. During the competition, whenever we were in trouble, we struggled together to work it out. As an international competition, the jury committee of ASC 2014 consisted of some of the most established experts in the domain, who set high standards and high requirements. Being focused and being positive were the two important suggestions I gave my students, and they did show a great performance on these two things all through the competition.

13.8 LICOM Optimization Summary and Evaluation

Looking back to the optimizations of LICOM, we thought that the early-term practice and experience about the configuration, and the optimization during the competition, were equally important. After practicing installing the NetCDF library many times, we found two practical methods to install it. Between the two, we chose the straightforward one, as introduced previously in Sect. 13.4; you needed to add "-lnetcdff" to the NLIB option in the case.sh script.

Despite the satisfying results achieved in the LICOM optimization, we understood that there were more optimizations that could have been done. Therefore, after the competition, we further used OpenMP to achieve MPI+OpenMP hybrid parallelization in LICOM2.0, and achieved a better performance.

At last, we would like to express our best gratitude to the organizing committee for providing us with the most advanced supercomputing platforms. We also wanted to thank the engineers from Inspur who offered us a lot of help and

Fig. 13.2 Group photo of TUT supercomputing team

guidance. We still had a lot to learn in terms of high performance computing, and we would always be ready to cooperate with other colleagues to step ahead. A group photo of TUT supercomputing team from Taiyuan University of Technology is shown in Fig. 13.2.

13.9 Further Reading

- **NETCDF C** See more information via http://www.unidata.ucar.edu/software/netcdf/.

Chapter 14
Optimization of Three Dimensional Elastic Wave Modeling Software 3D_EW

14.1 Introduction to the Team and the Preparation

14.1.1 Introduction to the Team

The 3D_EW (3D Elastic Wave Modeling) team of Shanghai Jiao Tong University (SJTU) in ASC14 contains two advisors, including Xinhua Lin, who is the deputy director of the High Performance Computing Center of SJTU, and Minhua Wen, who is an engineering specialist, and three students that are in charge of the work of code porting and optimizations. The backgrounds of the three students are as follows:

Jiaming Zhao, a senior student from the Department of Computing Science, is in charge of the code porting of 3D_EW to MIC and related optimizations.
Ye Yuan, a junior student from the Department of Computer Science and is in charge of the MPI implementation of 3D_EW.
Di Xiao, a junior student from the School of Software Engineering, is in charge of the profiling work using Intel Vtune and so on to provide technical support for other team members.

The 3D_EW team have the full support of the High Performance Computing Center: besides the MIC node provided by Inspur, they also have access to the 5 MIC nodes from the π cluster of the computing center. Each node contains two Intel Xeon E5-2670 CPU plus two Intel Xeon Phi 5110P, which are connected through 56 G IB FDR.

ASC Community, *The Student Supercomputer Challenge Guide*,
https://doi.org/10.1007/978-981-10-3731-3_14

14.1.2 Technical Preparation

As 85% of the computing power of Tianhe-2 comes from the KNC (Knights Corner) MIC accelerators, we need to study the KNC carefully before optimizing the 3D_EW program based on Tianhe-2.

Compared with CPU, KNC has three characteristics that can be considered for optimization.

14.1.2.1 512-Bit Vector Processing Unit (VPU)

(1) **VPU is the key factor for the performance of KNC**
The 512-bit VPU can perform 8 64-bit double-precision floating-point operations or 16 32-bit single-precision operations. The theoretical peak performance of KNC is: Peak Performance = Clock Frequency * Cores * Lanes * 2 FMA, where FMA (Fused Multiply and Add) means performing one floating-point multiplication and one floating-point addition in one cycle. Take the KNC 5110p from the π computer for example. The double-precision and single-precision performances are 1.05 GHz * 60 cores * 8 Lanes * 2 FMA = 1008 Gflops and 2016 Gflops, respectively. Therefore, to achieve better performance, we need to make full use of the VPU.

(2) **Unaligned data would reduce the efficiency of VPU**
Each thread in the KNC has 32 registers, each of which has a length of 512 bits. Therefore, it only takes one cycle to read aligned data, while two cycles to read not aligned data, which is the read operation.

(3) **Gather/scatter is the bottleneck for VPU**
KNC uses the IMCI-512 instruction set that supports the gather/scatter operations, which may easily become a bottleneck. The reason is that gather operation is implemented by reading each distributed Cache Line (CL) one by one. Therefore, the more CLs to read, the longer the time delay.

In summary, we need to avoid gather/scatter operations, as the 512-bit VPU has strict requirements for data alignment.

14.1.2.2 The Hardware Cannot Hide Cache Latency

(1) **High cache latency**
We use the KNC nodes of the π computer as an example. As shown in Fig. 14.1, the latency for the L1, L2, and L3 cache of Intel Sandy Bridge E5-2670 is 1.5, 4.6, and 81 ns, respectively, while the latency for the L1, L2 cache, and the memory of KNC 5110p is 2.9, 22.9, and 295 ns, respectively.

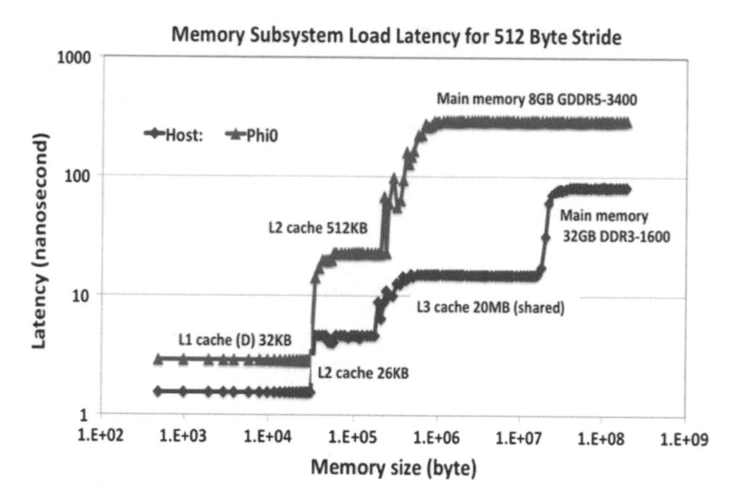

Fig. 14.1 Cache latency comparison between KNC and CPU

(2) **Low ILP**
Based on the empirical equation, Power consumption = (Performance) 1.73, the power consumption of P6 that uses Out-of-Order (OoO) is three times higher than P54C that uses in-order core. However, the performance is only increased by 1.9 times. Restricted to the power consumption and the technology, KNC uses P54C at the cost of a low Instruction Level Parallelism (ILP).

(3) **Less hardware cache prefetchers**
Restricted by the power consumption and the space, KNC does not have any L1 cache prefetchers. Even for L2 cache, it only has 16 stream prefetchers, and no stride prefetchers.

In summary, compared with KNC, a CPU can hide the cache latency by methods such as OoO execution and cache prefetchers. Therefore, we need to pay more attention for the cache prefetch when using KNC.

14.1.2.3 Limited Memory Bandwidth and Cache Size

(1) **Measured bandwidth is less than half of the theoretical bandwidth**
KNC has eight memory controllers, each of which has two 5.5 T/s channels. The size of each memory Transaction is 4B, so that the theoretical memory bandwidth should be 8 * 2 * 5.5 * 4 = 352 GB/s. However, Fig. 14.2 shows that the measured bandwidth using STREAM is only 140 GB/s, less than half of the theoretical bandwidth. On average, the bandwidth for each thread is only 0.5 GB/s, which is much less than that for the CPU.

(2) **Slow false sharing and remote access for L2 cache**
Each KNC 5110p core has 512 KB L2 cache, so that the total size is 30 MB. However, each core cannot directly access the CL of the L2 cache of other

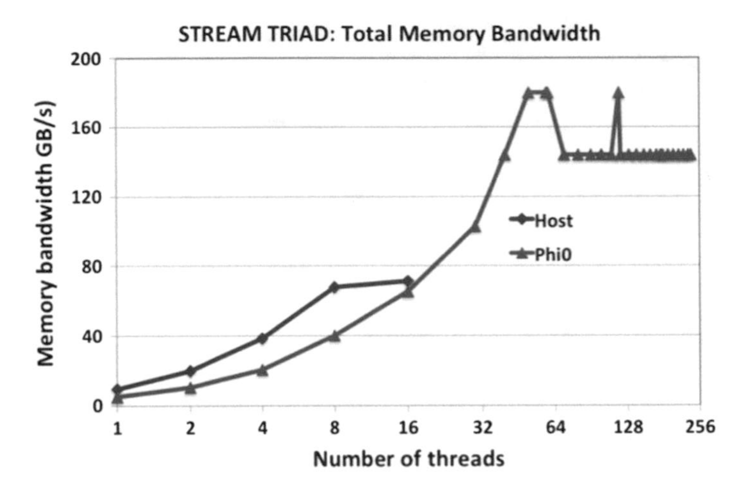

Fig. 14.2 Measured memory bandwidth for KNC

cores, and has to copy the CL to its own L2 cache. Furthermore, the speed of accessing other L2 cache is only 17% faster than accessing the main memory, and has no relationship with the physical distance between the two cores in the Bi-direction ring.

(3) **No L3 cache**

KNC does not have L3 cache because there is no extra space. Based on practical experience, the size of the next level cache should be eight times bigger than the current level caches. For example, the size of L1 cache for KNC is 32 KB (data) + 32 KB (instruction) = 64 KB. The size of L2 cache is 64 KB * 8 = 512 KB. Similarly, the size of L3 cache should be (64 KB + 512 KB) * 8 = 4.5 MB, and the size of the aggregated L3 cache should be 270 MB. Such a big cache size would require an area of at least 500 mm^2 even when using the 14-nm technology for the next-generation Intel Knights Landing. In contrast, the largest area of current Intel chips is only 200 mm^2.

In summary, due to the limitation of the memory bandwidth and the cache size, we should avoid communication between the cache and the memory, and use NT-store optimizations.

14.2 Explanation of the ASC14 3D_EW Rules

3D_EW is the most challenging topic in the ASC14. All teams are required to optimize the real 3D_EW application on the world's most powerful supercomputer, Tianhe-2. Tianhe-2 contains both tradition CPU and the MIC accelerators released in recent years by Intel. Therefore, we need to consider not only the intra-node parallelism through OpenMP, but also the inter-node parallelism through MPI. We also need to consider the parallelism through heterogeneous computing, namely the CPU-GPU computing. The parallel ideas include shared memory, massage passing as well as heterogeneous computing. It is a little difficult for undergraduate students.

3D_EW accounts for 20 points and is the largest part of ASC14. Unlike traditional competitions, there are 17 points for the regular session and 3 points for the final session. The regular session is open to all teams, while only the first two teams can attend the finals. Therefore, teams entering the finals will have advantages over other teams.

In summary, from both the perspectives of difficulty and importance, 3D_EW is the key task for ASC14. So our team put the most effort in on this topic.

14.3 Introduction to 3D_EW

3D_EW is a real application used in oil exploration and is developed by BGP Inc of China National Petroleum Corporation. 3D_EW uses the wavefield downward continuation method to simulate the propagation of elastic wave in an elastic homogeneous medium. In the program, the vertical wave (P wave) and the horizontal wave (S wave) are simulated separately to better understand their propagation. The propagation can be simulated using the high order finite difference method. The wavefield equation is written as:

$$\rho \frac{\partial^2 \mathbf{S}}{\partial t^2} = (\lambda + \mu)\nabla\theta + \mu\nabla^2 \mathbf{S}$$

Figure 14.3 shows the serial program of the algorithm. After the initialization of input data, the program steps into the ishot loop, which is to loop among different shots. The ishot loop is independent and can be easily parallelized. Loop l is the time step loop, and the data in each step is related to its previous step. Within the loop l is the loop among different mesh points. Based on the explicit finite difference, the computing of each point is completely independent and is also easy to parallelize.

14.4 Analysis of the 3D_EW Program

14.4.1 Hotspot Analysis

The structure of 3D_EW does not involve any complicated function executions. The main computation happens within the loop. There is data dependency between different iterations of loop 1. Loop l includes three levels of loops (k, j, i), as shown

```
initialize the wave field in the 3D domain
for(ishot=0;ishot<nshot;ishot+1)
//different ishot means different source of ware
{
......
    for(l=0;l<lt;l++)//lt means time step
    {
    ......
            for i,j,k in x,y,z direction
            {
                    compute P-wave and S-wave in each grid of the 3D domain
            }
    }
}
```

Fig. 14.3 3D_EW program structure

```
ntop = ntop-1;
nfront = nfront-1;
nleft = nleft-1;
for(k=ntop;k<nbottom;k++)
    for(j=nfront;j<nback;j++)
        for(i=nleft;i<nright;i++)
        {
            if(i==ncx_shot-1&&j==ncy_shot-1&&k==ncz_shot-1)
            {
                px=1.;
                sx=0.;
            }
```

Fig. 14.4 The *kji* loop of the 3D_EW programs

in Fig. 14.4. The lowest loop *i* contains only addition and multiplication operations. Obviously, the hotspot is the *kji* loop.

14.4.2 Code Analysis

We found three characteristics after analyzing the 3D_EW program:

Easy to parallelize There is no data dependency for both the ishot loop and the *kji* loop. So the whole program is easy to parallelize.

Unit stride access It is a large number of Unit Stride Access for memory access, in the *kji* loop, we need to read the five neighboring points to compute each mesh point. The locality on *x* direction is very good, but the points on the *y*, *z* directions are among different columns and different slides. Such data accessing pattern is not efficient especially for a KNC platform, which cannot hide the cache latency, as we discussed before.

Regular data structure and balanced loads This data structure has a great influence on the performance in high performance computing. The data used in 3D_EW has simple data structures, which is very suitable for parallel computing. Besides, the computing domain is a simple 3D topological structure, and there is seldom a loop branch. Even though the computing domain is increased with the value of l, the amount of computation tends to be stable when the domain reaches the settled boundary. The amount of computation is almost identical for *kji* loop so that the whole program has a very good load balance.

14.5 Parallel Design and Implementation of 3D_EW on the Hybrid CPU + MIC Cluster

14.5.1 The General Parallel Design for 3D_EW Based on the Hybrid CPU + MIC Cluster

As 3D_EW has a huge amount of data to compute, we use MPI + OpenMP to implement the parallel design based on multiple heterogeneous nodes. MPI is used to implement the inter-node parallelism. As the ishot loop is completely independent, we assign them among the MPI processes and different compute nodes. The OpenMP is used to implement the inner-node parallelism and the SIMD vectorization. Loop l has data dependency, and cannot be fully parallelized. But the *kji* loop has no dependency so we use OpenMP to parallelize it.

14.5.2 Implementation of 3D_EW Based on CPU + MIC Cluster

14.5.2.1 OpenMP Multi-threading

OpenMP is a parallel programming standard based on shared memory architecture. We can use OpenMP to implement the multi threading computing based on a multi-core system, and to achieve parallel acceleration. Compared with parallel computing based on multiple processes, OpenMP is a lightweight parallelization approach, and all threads share one memory space. Therefore, the expense of data sharing is very small and it is very suitable for intra-node parallelism.

As the ishot loop also has no data dependency, technically we can also use OpenMP to parallelize such a loop. However, we use MPI instead for two reasons: (1) we have multiple compute nodes; and (2) we need to create an isolated space for different ishots, which have a heavy burden on the memory. We will introduce the technology of MPI later. We ignore parallelizing loop 1 as it is not suitable. For the *kji* loop, which is also easy to parallelize and contains the largest computation, using OpenMP is more suitable. The *k* loop is the most time-consuming part in the *kji* loop. However, we choose not to parallelize it as each loop is only executed five times and both multi-core CPU and many-core MIC are not suitable for use here. Therefore, here we use OpenMP to parallelize the *kji* loop.

As for the *kji* loop, since we need to further optimize the inner-loop (using SIMD vectorization), we put the OpenMP on the outer-level loop, namely the *k* loop. Besides being a lightweight parallelization method, OpenMP is also easy to use. Programmers only need to add some compilation directives before the corresponding loop.

Figure 14.5 shows the way to use the OpenMP multi-threading method.

14.5.2.2 SIMD Data-Level Parallelism

SIMD (Single instruction, multiple data) is a data-level parallelism that employs a controller to control many micro units and apply the same operation on multiple data items in one group. Many current processors with cache structures have support for SIMD instructions. In the processors we use, the Sandy Bridge CPU has a 256-bit VPU, while MIC has a 512-bit VPU. SIMD greatly improves the performance of parallel computing. For example, the 512 bit MIC VPU can process 8 double precision or 16 single precision data simultaneously.

We apply SIMD on both the CPU and MIC parts when optimizing 3D_EW. Both CPU and MIC are based on the ×86 architecture so that the vectorization methods are similar, with adding "#pragma simd" before the target code.

As the VPU is based on the same cache system, more intensive data read and more regular computing patterns would generally bring better performance. Most data in 3D_EW is three dimensional and stored in line sequence in C program. The inner index of array elements for *kji* loop is also based on *i*. So we implement the vectorization inside the *i* loop, as shown in Fig. 14.6. In this way, the data stride for

```
#pragma omp parallel for default(shared) private(px, k, i, j, kk, kkk, mvvp2, mvvs2, mtempux2, mtempuy2, \
        mtempuz2, mtempvx2, mtempvy2, mtempvz2, mtempwx2, mtempwy2, mtempwz2, mtempuxz, mtempuxy, mtempvyz, \
        mtempvxy, mtempwxz, mtempwyz) num_threads(TCPU)
        for(k=ntop;k<nbottom;k++)
        {
                if(k + zlow - 5 <205)
                {
                    mvvp2 = 2300 * 2300;
                    mvvs2 = 1232 * 1232;
                }
```

Fig. 14.5 OpenMP multithreading

```
                for(k=ntop;k<nbottom;k++)
                {
                    for(j=nfront;j<nback;j++)
#pragma simd
                        for(i=nleft;i<nright;i++)
                        {
                            for(kk=1;kk<=mm;kk++)
                            {
                                for(kkk=1;kkk<=mm;kkk++)
                                {
```

Fig. 14.6 SIMD data-level parallelism

```
..LN2796:
   .loc    1  256  is_stmt 1
        vscatterpf0hintdpd (%r14,%zmm0,8){%k1}                          #256.21 c13
..LN2797:
        vscatterpf0hintdpd (%r14,%zmm0,8){%k1}                          #256.21
..LN2798:
        nop                                                             #256.21
..L300:                                                                 #
..LN2799:
        vscatterdpd %zmm2, (%r14,%zmm0,8){%k1}                          #256.21
..LN2800:
        jkzd        ..L299, %k1    # Prob 50%                           #256.21
..LN2801:
        vscatterdpd %zmm2, (%r14,%zmm0,8){%k1}                          #256.21
```

Fig. 14.7 Large number of gather/scatter operations in the assembly code

the same SIMD is decreased so as to achieve better locality and more efficient utilization of the read/write bandwidth.

Using SIMD can achieve better performance on CPU than that on KNC, as KNC cannot hide the high data read latency. As for 3D_EW, the hotspot involves a large number of Unit Stride Accesses, which means the data access is very distributed. Therefore, after vectorization, VPU will have a lot of inefficient gather/scatter operations when accessing data (shown as the assembly code in Fig. 14.7). In the SIMD implementation on MIC, we can use low-level C Intrinsic (as shown in Fig. 14.8) to control the data access operations, so as to decrease the gather/scatter operations, and to achieve more performance improvement.

Besides, we also align the data when implementing the SIMD vectorization. Using mm_malloc to allocate memory on MIC can guarantee the data alignment and read the Cache Line into the register in one cycle, as shown in Fig. 14.9.

If the data is unaligned, we need two cycles to read one CL into the register, as shown in Fig. 14.10.

```
_a = _mm512_load_pd(mic0_wp1 + k*ny*nx+j*nx+i);
_b = _mm512_load_pd(mic0_wp2 + k*ny*nx+j*nx+i);
_c = _mm512_add_pd(_mtempuxz, _mtempvyz);
_c = _mm512_mul_pd(_tem, _c); _c = _mm512_add_pd(_c, _mtempwz2);
_a = _mm512_mul_pd(_arrayc, _a); _c = _mm512_add_pd(_a, _c);
_c = _mm512_sub_pd(_c, _b);
_tem = _mm512_load_pd(pxtem);
_c = _mm512_add_pd(_c, _tem);
_mm512_mask_store_pd(mic0_wp + k*ny*nx+j*nx+i, _mask, _c);
```

Fig. 14.8 C Intrinsic optimization

```
_a = _mm512_load_pd(mic2_vp1 + k*ny*nx+j*nx+i);
```

Fig. 14.9 Aligned data access

```
_a = _mm512_loadunpacklo_pd(_a, mic2_u+ k*ny*nx+(j+kkk)*nx+(i+kk));
_a = _mm512_loadunpackhi_pd(_a, mic2_u+ k*ny*nx+(j+kkk)*nx+(i+kk+8));
```

Fig. 14.10 Unaligned data access

14.5.2.3 CPU-MIC Cooperative Computing

As the performance of one CPU is 0.3–0.4 TFLOPS, the computing power for the two CPUs inside one node is also comparable to the 1 TFLOPS of the MIC card. As mentioned above, the *kji* loop is easily parallelized, so we could assign part of the computing task to the CPU. Besides, some cases provided by ASC have large memory requirements beyond the 8 GB MIC memory, so that we have to decompose the computing domain and use multiple MICs. Here we simply do the 1-dimensional decomposition on the z direction as data is stored in the sequence of x, y, and z. Such a domain decomposition is shown in Fig. 14.11, where the task allocation for CPU and MIC is based on their computing power. Data will be exchanged between neighboring cards for the next time step computation.

14.5.2.4 MPI Implementation

MPI (Message Passing Interface) is a parallel programming standard based on massage passing and is supported by most parallel computers. MPI has good portability. We can use MPI to create multiple processes and assign the coarse-grained computing tasks among different processes. MPI processes do not

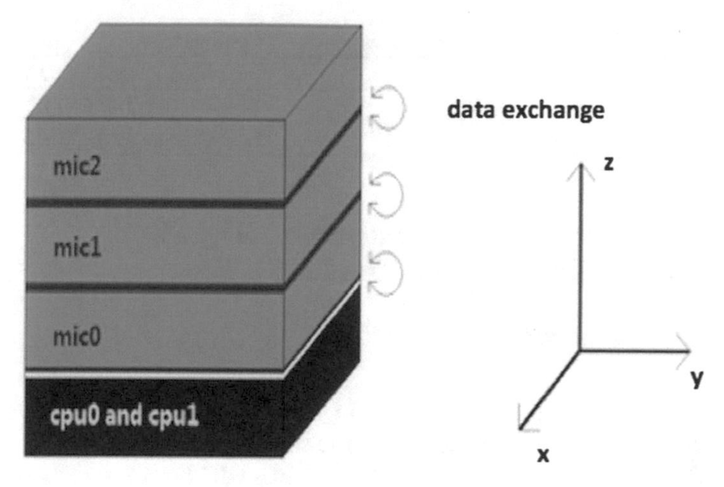

Fig. 14.11 Data exchange between CPU and MIC within one node

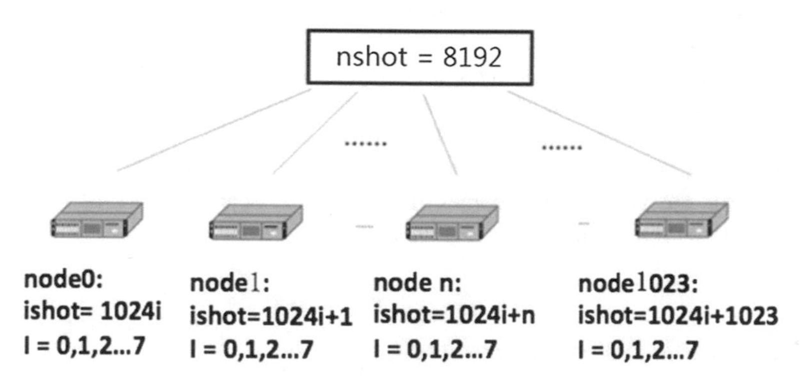

Fig. 14.12 Allocating the tasks on 1024 nodes

have shared memory, so that the data has to be exchanged through massage passing. Thus MPI has high cost for shared-memory programming.

As the ishot loop does not have data dependency and is in the outer level, MPI inter-node parallelism is very suitable. When the scale of the ishot is very large we can use MPI to assign different ishots into different processes, as shown in Fig. 14.12, where the 8192 shots are allocated between 1024 compute nodes.

Table 14.1 Single-node configuration of π cluster

Software/hardware	Configuration
CPU	Intel SNB E5-2670 @ 2.7 GHz
KNC	Intel KNC 5110P @ 1.05 GHz
Compiler	Intel Composers XE 2013

14.6 The Experiment and Analysis of 3D_EW Optimization Based on Tianhe-2

14.6.1 Environment of the Platform

The parallel design and optimization of 3D_EW is mainly based on the π cluster (small scale), while the large-scale test in the final is based on Tianhe-2.

Developed by Inspur, Π cluster is a CPU-GPU-MIC heterogeneous supercomputer deployed in the High Performance Computing Center of Shanghai Jiao Tong University. The peak performance is 263 TFLOPS, and the HPL measured performance is 192 TFLOPS. π is for now the most powerful supercomputer in the universities administrated by the Ministry of Education. Table 14.1 shows the single-node configuration:

Tianhe-2 is a self-designed heterogeneous supercomputer with a peak performance of 53.9 PFLOPS. The HPL measure performance is 33.0 PFLOPS in double precision, which has been first in the Top500 list three times. Tianhe-2 has 16,000 compute nodes, with 32,000 CPU processors, and 48,000 MIC accelerators. More than 85% of the performance comes from the MIC cards. Table 14.2 lists the single-node configuration.

14.6.2 Single-Node Performance

We use two cases to test the single-node performance, as shown in Figs. 14.13 and 14.14, where:

- the first bar refers to the serial program;
- the second bar refers to the speedup after using OpenMP CPU;
- the third bar refers to the further speedup using SIMD on CPU;
- the fourth bar refers to the speedup on CPU + MIC cooperative platform, without using C Intrinsic; and
- the fifth bar refers to the speedup with the MIC part using C Intrinsic.

As for small-scale test, including MIC will decrease the performance, as more time was spent in the data exchange between CPU and GPU.

As for the medium-scale cases, involving MIC for computing can bring a slight performance improvement, as MIC cannot achieve good computing performance.

Table 14.2 Single-node configuration of Tianhe-2

Software/hardware	Configuration
CPU	Intel IVB E5-2692 @ 2.2 GHz
KNC	Intel KNC 31S1P @ 1.1 GHz
Compiler	Intel Composers XE 2014

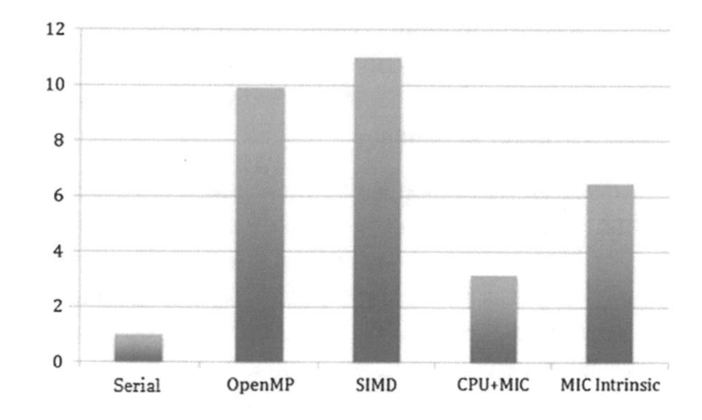

Fig. 14.13 Speedup in small-scale tests

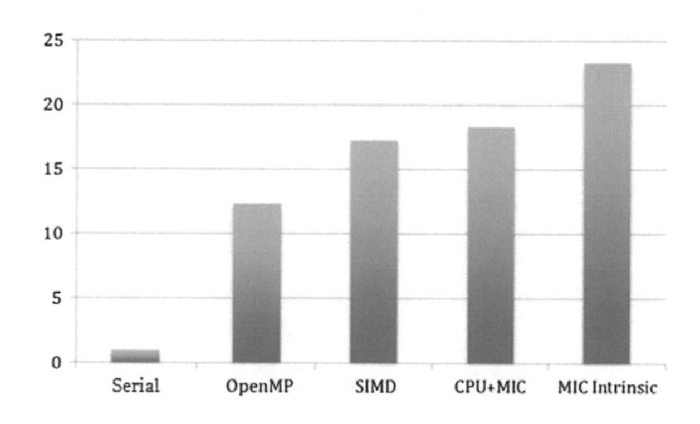

Fig. 14.14 The speedup in the medium-scale tests

The MIC performance is greatly improved after using C Intrinsic optimizations. The reason has been discussed previously.

14.6.3 The Performance Based on Inter-node Platform

In the final test based on Tianhe-2, the ishot is set to 8192 and we have used 1024 compute nodes, 1/16 of the total. Therefore, each compute node is in charge of nine ishots. As there is no data dependency between different ishots, we can easily achieve good scalability and linear speedup.

14.7 Experience and Closing Remarks

ASC is one of the three world-famous supercomputing competitions, and it is the second time our university has participated. Our target is to improve our standards in the preliminary stage and the final stage, and insist on the following ideas:

- To bring more students into the field of supercomputing, and to foster more talent for our university through this competition.
- To learn the most advanced knowledge from our brother universities and to improve our experience.

Therefore, we stay positive throughout the competition. As for myself, the most meaningful achievement is not only the final championship, but also the five students with talent and passion for the supercomputing domain. The students showed a lot of interests in supercomputing, and were able to find and solve the problems by themselves. Especially in the 3D_EW optimization, as Tianhe-2 is the most powerful supercomputer, we have studied a lot to understand its architecture and put in a lot of effort in the optimization. Optimizing the 3D_EW application strengthens our capabilities and experience of the MIC platform. As for the students, the competition has a great positive influence on all of them. Most students stay in our High Performance Computing Center for further researches. One student is now doing his Ph.D. studies in the USA.

Printed in the United States
By Bookmasters